For

John

With admiration, blessings

and the best good wishes.

Ed

*Harvesting the Air*

The earliest depiction of a post-mill, an initial letter
from a copy of Aristotle's *Meteorologica*, painted in
England about 1250–1260. From Harley MS. 3487, fol. 161;
reproduced by permission of the British Library. About the
same time, another artist decorated a second copy of this
treatise with a remarkably similar windmill (see Figure 5).

# Harvesting the Air

## Windmill Pioneers in Twelfth-Century England

EDWARD J. KEALEY

University
of California
Press

BERKELEY · LOS ANGELES

University of California Press
Berkeley and Los Angeles, California

© 1987 by
The Regents of
the University of California

LIBRARY OF CONGRESS CATALOGING-IN-PUBLICATION DATA
Kealey, Edward J.
    Harvesting the air.
    Bibliography: p.
    Includes index.
        1. Windmills—England—History. I. Title.
TJ823.K43   1986   621.4′5   85-24627
ISBN 0-520-05680-9 (alk. paper)

Printed in the United States of America
1   2   3   4   5   6   7   8   9

*T*o the memory of my beloved Mother,
MARGARET L. KEALEY,
who died as this book was being completed,
and to her four grandchildren,
KERRI and TOM,
DONALD and DOREEN,
our hope in a new generation.

# Contents

# Illustrations

# *Preface*

---

*T*his study explores the social context of a technological revolution. It traces the origin of the Western windmill to the early twelfth century and outlines that invention's contradictory effects. A cavalcade of fascinating, forgotten personalities championed, exploited, and even opposed wind-powered machines. These individuals once debated everything from ecclesiastical issues to land reclamation. Now they can introduce readers to four generations of surprising change.

Not long ago there were millions of windmills, but most disappeared with the spread of electricity. Some countries lost them altogether. The remnants were thought to be quaint decaying landmarks, until the energy crisis of 1973. Then people learned that thousands of bent blades had never stopped turning and realized that their faithful productivity offered one viable solution to current energy needs. Soon countless new windmills were springing up all over.

Coincidentally, my research in Anglo-Norman documents uncovered exciting new information about the earliest mills and the men and women who promoted them. Some of the characters were generous and some were devious, but all optimistically encouraged commercial enterprise and a wide dispersal of labor-saving techniques. Their farsighted, typically medieval, trait of employing basic knowledge to enhance the general welfare was no accident, for mechanical skill was a valued asset on the pilgrim road to eternity. Humility was another pilgrim virtue, but it effectively prevented the earliest wind-power pioneers from gaining much recognition in their own time. It is pleasing and appropriate to salute them now when everyone is seeking energy independence.

Many scholars have cheerfully aided my investigation. Dr. Mary Cheney of Lucy Cavendish College, Cambridge University, and Professor Lynn White, Jr., of the University of California at Los Angeles, offered stimulating commentary at an early stage of my search. Professor William Reedy, of the State University of New York at Albany, loaned microfilms of monastic cartularies; Professor Ellen Kosmer, of Worcester State College, volunteered several significant references; and Professor Alfred Desautels, S.J., of the College of the Holy Cross, translated obscure passages. Professor A. H. de Oliveira Marques, of the University of Lisbon, assisted me with Iberian precedents, and Professor Christopher Cheney, of Corpus Christi College, Cambridge University, supplied needed details on Canterbury matters. Dr. Marjorie Chibnall, of Clare Hall, Cambridge University, and Professor C. Warren Hollister, of the University of California at Santa Barbara, encouraged me in numerous ways. I owe a special debt to my friends from the "Painter School" of English Medieval History at The Johns Hopkins University, 1958–1962; we have happily traded ideas and mutual support for a quarter-century. The talented editors at the University of California Press also graciously enhanced my presentation.

Archivists at many record depositories, particularly at the Bodleian Library, the British Library, the Cambridge University

Library, the Canterbury Cathedral Archives, the King's College Library (Cambridge), the Magdalene College Library (Oxford), and the Public Record Office courteously answered unusual questions and requests for more and more charters. For special help with the photographs I am indebted to Dr. Adelaide Bennett and Dr. Nigel Morgan, of the Index of Christian Art, Princeton University; C. M. Hall of the British Library Department of Manuscripts; Arthur Owen, Keeper of Manuscripts, Cambridge University Library; Professor Henry Loyn, of the University of London; P. S. Jarvis; Anne and Scott MacGregor; Violet Pritchard; Terence Paul Smith; David and Mark Sorrell; and Robert J. Zeepvat, of the Milton Keynes Archaeological Unit.

The Penrose Fund of the American Philosophical Society supported my research during the summer of 1980, and the College of the Holy Cross awarded me a fellowship in the summer of 1982, a sabbatical in 1983–1984, and a fellowship in the spring of 1986.

Near day's end, good-hearted millers would usually set their sails at special angles to the horizon, thereby bidding all viewers peace and joy. May the following pages convey a similar message.

# Searching the Sky

*E*uropean entrepreneurs were the first people in history who consistently combined technological expertise with natural energy to create labor-saving devices. Their special pride, and their unique problem, was the wind-powered post-mill. It complemented the older water mill and enlarged medieval society's unprecedented attempt to make life better for everyone. A strikingly innovative machine in a world of handicraft skills, it attracted numerous imaginative, sometimes contentious, supporters. Although nourished by the same traditions that had fostered chivalric knighthood and courtly love, the new invention encouraged average people to challenge baronial authority. From the beginning the windmill confounded, as well as captivated, its champions.

Appealing, productive, and even mysterious, the windmill was above all a triumph of ingenuity over toil. Its quartered sheets spun with unseen cause, but its shape and utility were instantly recognizable. Nevertheless, this dynamo stretched its infant arms without ceremony or fanfare. Although subsequent observers would enthusiastically sing its praises and paint its romance, the earliest bards, chroniclers, and engineers who contemplated the wondrous apparatus left few personal recollections. Their indifferent silence has too long concealed the emergence of one of humanity's most altruistic creations.

Even now, the haunting breezes still whisper, "When and where did the windmill first catch breath, and how were people originally affected?"

The quest for such hidden knowledge is beset with false trails and episodic detours. Accordingly, it seems best to report the principal conclusions at the outset and then to trace the engrossing byways. Boiled down to the small, my findings indicate that the Western windmill was perfected in England and that it initially embraced the air sometime before 1137. This is half a world away from, and half a century earlier than, the conclusions of previous investigators.

Scholars could once identify only a handful of early post-mills, but I have located fifty-six examples in England alone before 1200. Most of the new facilities were discovered in prosaic sources—the overlooked financial records of ambitious Anglo-Norman landlords. These documents demonstrate that some customers, operators, and patrons encountered problems in addition to enjoying profit. Fortunately, their minor distress was far outweighed by a vast improvement in the lives of countless persons throughout succeeding centuries.

This study of the earliest windmills will proceed in three stages. First, the significance of the windmill's possible ancestors, especially the vertical Roman water mill and the horizontal Persian wind tunnel, will be briefly outlined. Then the actual design of the Western post-mill will be examined. Finally, and most important, extant twelfth-century records will be scrutinized so that the critical achievements of the first wind-power enthusiasts can be declared.

From time immemorial hardworking farmers and leisured visionaries dreamed of taming the boundless powers of nature, but their early accomplishments were decidedly trivial. Greek and

Roman scientists occasionally tested the unknown forces within earth, air, fire, and water, but they were usually motivated only by isolated curiosity or frivolous pleasure. Perhaps it was a superstitious awe of the divine spirits which supposedly animated all things that made ancient technicians so hesitant to exploit nature's full potential. An intellectual contempt for manual labor, a pervasive reliance on slave workers, and a lack of farsighted investors may have further inhibited practical technological inquiry.

There was, however, a single striking exception to such classical disengagement. One hundred years before the birth of Christ, someone in Europe devised the water mill—a revolutionary event, since this was the first truly complex machine. Contemporaries quickly grasped its significance, for the rate of power thus bridled was roughly equivalent to the grinding effort of fifty women. About 85 B.C. a Greek poet named Antipater trumpeted the glad news, urging tired maids who labored at hand mills to rest because now veritable water nymphs would "throw themselves on the wheel, force round the axle tree, and so the heavy mill." His ode is one of the most convincing proofs that a new day had dawned in power technology.[1]

The water mill soon began to appear in other widely scattered places: at a palace of Mithridates, king of Pontus, in 63 B.C., where it amazed Pompey's conquering troops; in China in A.D. 31, where it operated a bellows; and in Jutland in the first century A.D., where excavated water channels suggest its probable existence.[2] Such gristmills were most likely all of the type later called

1. Quoted in Richard Bennett and John Elton, *History of Cornmilling*, 2:6. There were three contemporary poets named Antipater, but the Antipater of Thessalonica is most likely the one referred to here.

2. Terry S. Reynolds, *Stronger than a Hundred Men: A History of the Vertical Water Wheel*, is a superb new history of the water mill. For its early history, see pp. 1–120. On p. 12, Reynolds provides a slightly different translation of Antipater's poem. Even earlier than water mills were the primitive Eastern *noira*, revolving wooden rims with attached bailing buckets that raised stream water from one level to another. Since noira performed no other function, they are not considered water mills, nor is their lineal relationship to true mills clear.

the Greek (or Norse) mill, whose grinding stones were driven directly from below as water struck the attached blades of a submerged spindle. No gearing was involved. This design is also called the horizontal water mill, because of the plane in which the blade wheel rotated.

A second type of water mill featured a large side wheel that spun an axle and attached gears. Because the wheel turned in a vertical plane, this "Roman" design has been classified as a vertical mill. It was also called a Vitruvian water mill after the architect and engineer Marcus Vitruvius Pollio, who described it in his treatise on architecture, apparently written during the principate of Augustus (27 B.C. to A.D. 14). River water could activate the Roman mills by striking their paddle wheels from beneath, or in an overshot arrangement, which allowed water to be dumped on the blades from above.

Unfortunately, the water mill was rarely utilized to full capacity in its early days. Some Mediterranean areas lacked sufficient water power, but even when it was available, classical authorities usually preferred to rely on cheap slave labor, rather than to put technical knowledge at the service of popular needs. It would be about five hundred years before the water mill would be properly employed.

One isolated example illustrates what was accomplished when the water mill was finally used. In the early fourth century A.D., the Romans erected a gigantic complex of sixteen waterwheels at Barbegal, near Arles in Provence. The connected flour mills were built on the slope of a hill in descending order and were fed by a long aqueduct. There were eight tiers with two mills on each level. The cascading water probably turned vertical overshot wheels.

Estimates suggest that this single installation could process flour for eighty thousand people, a considerable achievement. Arles was an imperial capital at the time, and the mills may have serviced divisions of the Roman army as well as the local populace. Although the Barbegal station did not become a model for other large facilities, several smaller individual water mills have been found near legionary posts along Hadrian's Wall in Britain. Near Mainz there was a large water-driven sawmill complex for cutting marble.[3]

Sometimes mills were put to sinister uses. About 290, when the Arles complex was being planned, a dreadful persecution of Christians launched by the Emperor Maximian was in full swing. One of its victims, Saint Victor of Marseilles, was reportedly crushed beneath the grist stone of a mill and then beheaded. His tomb became one of the most popular pilgrimage centers in France. Artistic representations of the martyr usually depicted him beside a millstone, although over the ages his iconography changed, and he was sometimes anachronistically represented with a windmill sail or a full post-mill.

The Romans and other classical peoples clearly possessed the expertise to convert water power into productive energy but evidently lacked the inclination to make hydraulics widely available. Those who did utilize water power seem to have been governmental officials or well-financed industrialists. Despite the poet's song, most grain was still laboriously ground by hand or milled by animal tread.

Only in the Middle Ages did people really begin to pursue a deliberate policy of mechanization. Christian thinkers came to detest slavery and to emphasize the dignity and welfare of all individuals. Socially minded European experimenters thereafter consistently yoked natural energy to labor-saving devices. Such regular conversion of power into work was unprecedented, but

3. Stuart Fleming, "Gallic Waterpower: The Mills of Barbegal"; Reynolds, *Stronger than a Hundred Men*, pp. 38–44. See also Walter Horn and Ernest Born, eds., *The Plan of St. Gall*, 2:229, 308–16, and 215–48 for a discussion of the milling process.

medieval men and women quickly surrounded themselves with machines of all types.[4]

In particular, vertical water mills became the ubiquitous organs of productive power. Each settlement wanted its own. The primary function of these mills was always grinding grain (wheat or rye) into flour. However, there were also early drainage mills and malt mills, used for making beer mash. Medieval people even extended the basic machinery in ways their predecessors had never envisioned.

The vertical mills had internal gears to rotate their millstones, but a further modification eventually transformed simple rotary motion into reciprocating motion. The invention of the camshaft and its associated trip-hammer opened the possibility of totally new uses for vertical water mills. Soon they were pounding and forging iron, sawing wood, fulling cloth, separating hemp, and reducing bark for tanning. Northern France evidently experimented most creatively with the new uses of water power that were dependent on rotary motion; the Alpine regions seem to have developed most of the novel applications of the trip-hammer.[5]

The English were also extremely fond of engines. By 1086, the year of the nationwide Domesday survey, they were so involved in hydraulics that many parts of the kingdom could boast one sturdy water mill for every fifty families. Riverfront mills sometimes even numbered three to a mile. The Domesday Book listed 5,624 water mills in some 3,000 different locations, and the extant copy does not cover the whole country.[6] There were also treadmills for animals, age-old hand mills, tidal mills to exploit

---

4. This subject has recently received considerable attention; see Carlo M. Cipolla, *Before the Industrial Revolution*; and Jean Gimpel, *The Medieval Machine: The Industrial Revolution of the Middle Ages*. Bradford B. Blaine of Scripps College, California, is preparing a major study of the use of water power in medieval industry.

5. Reynolds, *Stronger than a Hundred Men*, pp. 95–96.

6. Bennett and Elton, in *History of Cornmilling*, 1:131–80, list all the water mills; also see Gimpel, *The Medieval Machine*, p. 12. For a description of a typical water mill, see Philip Rahtz and Daniel Bullough, "The Parts of an Anglo-Saxon Mill."

the sea, and probably bridge and boat mills as well. Another major increase in water mill construction seems to have taken place between 1150 and 1250, both in England and on the Continent.[7] Existing mills also moved about the landscape in search of deeper waters, higher falls, and stronger currents for greater power.

With all this technological inventiveness and enhanced milling capacity, it is no surprise to us now that an innovative wind-driven mill arched its arms across the sky early in the twelfth century. At that time, however, the invention of the windmill must have seemed utterly marvelous—literally something new on the horizon. What may be startling today is that it was English, rather than Continental, technicians who first tamed the whistling free air and cleverly exploited its immense commercial potential. In view of the previous Continental success in developing novel uses for the water mill, it seems only fitting that the British Isles should finally have a mechanical triumph of their own. New evidence could disprove this conclusion, but as of this writing, it stands firm.

The testimony as to England's achievement is found in quite commonplace sources. For a variety of reasons, medieval people were extraordinarily generous to religious and charitable institutions. Fortunately, many benefactors carefully announced their donations in public ceremonies and in parchment charters; references to them are also found in legal decisions. Conscientious, or conscience-stricken, magnates gave away mills of all types, usually stipulating that each such machine was not to compete with another. These endowments and their accompanying directives, many of which are still unpublished, constitute important, hitherto unrecognized evidence concerning the establishment and distribution of all mills, including the earliest windmills. They prove that twelfth-century Britons, unlike other peoples, were

7. Reynolds, *Stronger than a Hundred Men*, p. 53.

familiar not only with the swish of paddle wheels and the creak of wooden gears but also with the flap of canvas sails.

Technological progress in England was only one facet of the new mill experience. Social change was another. Windmills were cheaper to erect and could function under different conditions than their water-powered predecessors, but the realization that their source of power—the blowing air—was freely available to everyone was truly unsettling.

The prospect of an unlimited increase in the number of private windmills threatened both lucrative old water mill franchises and traditional upper-class privileges. The post-mill offered quick-witted peasants an opportunity to evade manorial regulations, act independently, and become quite prosperous. In apprehensive reaction, many of the knights who had first idealistically championed a wide dispersal of the novel technology and had considered windmills excellent charitable donations later selfishly fought to regain control of local competition and to monopolize mill construction.

As centuries passed, the irrepressible wind encountered ever more sophisticated machines. Windmills enabled Western cultures to attain incredibly high levels of agricultural production, especially by pumping water to vast semiarid lands similar to the Great Plains of North America. Yet as we know, the demands of industrialization and the development of steam, electricity, fossil fuel, and nuclear energy ultimately made wind and water power seem antiquated. Sadly, thousands of proud sentinel mills were allowed to crumble or became quaint tourist attractions. Like derelict lighthouses, they were only haunting symbols of a different era.

Some mills refused to surrender to the seductive pace of progress. By the mid-twentieth century, only a few dozen windmills in England continued to chase their hands backward around the clockface of the sky, but in other parts of the world—especially North America, Australia, and South Africa—the majority of them kept right on turning. In the United States, thousands of lofty, pylon-supported wheels still effortlessly pump underground water on farms and ranches and at supply depots. A limited number desalinate ocean water, or generate electricity, as did many during the Second World War. One of history's oddest contrasts was the existence of a windmill at Los Alamos in 1944 when the atomic bomb was being developed.

The old machines declined in number, but they endured. In 1973, the world energy crisis suddenly made solar, hydroelectric, geothermal, and wind power popular again. People rediscovered that such power is safe, locally derived, inexhaustible, and non-polluting. Like the broad sweeping sails of the old mills, the available resource alternatives have thus been buffeted full cycle. California currently has more windmills than any other place in the world. One financial analyst has even predicted that by the year 2000, Americans will be using eighteen million windmills for auxiliary power. Another estimate suggests that 20 percent of the electrical production in the Netherlands will be wind-generated by the end of the twentieth century.[8]

It is only a moderate leap backward in time from these astonishing contemporary projections to the antecedents of England's twelfth-century use of wind power. Sailboats, of course, are a prehistoric triumph, but terrestrial exploitation of the free air is a relatively recent phenomenon. In the first century A.D. a Greek naturalist, Hero of Alexandria, toyed with a wind-driven pipe organ and left a careful description of his model. However, the earliest trace of a truly productive wind-powered mill appears much later, in the East.

In the tenth century Arab travelers noted that in Seistan, located in what is now the Baluchistan area that borders Afghanistan and Iran, a wind blew constantly from one direction, effort-

8. *New York Times*, April 27, 1980, and August 17, 1980. Other estimates vary widely.

lessly turning gristmills and raising irrigation water.[9] Today, in summer, the howling air sometimes races across the forbidding landscape for 120 days at a stretch. Consequently, the Persian wind machine had to be a durable creation.

Rather squarish, it was essentially a two-story mud brick tower that faced directly into the oncoming wind. The constant breezes blew through a gap in the upper front wall and emerged through an opposite hole. In their swift passage across the enclosed space, they spun a vertical spindle that had lightweight reed vanes mounted along its full length. The resulting rotary motion was then transmitted to grinding or pumping facilities on a lower floor. Usually several of these sturdy mills (which were really wind tunnels) were built right next to each other, like attached row houses. Their descendants are still in use today. The engineering concept is remarkably like that of the Greek water mill, for its vanes spin in a horizontal plane.[10]

Seistan is just south of the great caravan road leading to India and the Far East. Undoubtedly adoption of the new power engineering largely followed the direction of the highway, but spread of the technology may not have been very rapid. China and Tibet had long used somewhat similar, but not necessarily related, techniques to make small prayer wheels perpetually revolve. Functional upright wind machines appear in Oriental records only in the mid-thirteenth century, at the beginning of the Mongol pe-

9. In the tenth century, the Baghdad-born geographer al-Mas'udi noted that inhabitants of Seistan were renowned for their ingenuity in employing the wind to turn mills. His contemporary, ibn Haukal, also mentioned the existence of wind-powered mills in Seistan. See Hugh P. Vowles, "An Inquiry into the Origins of the Windmill," pp. 7–8. Furthermore, al-Mas'udi and another writer, at-Tabari, reported that Caliph Omar I (634–644) had commanded a Persian slave to build a mill operated by the wind and that the desperate slave assassinated the ruler. Obviously, this tale does not confirm the seventh-century existence of a windmill; see Lynn White, Jr., *Medieval Technology and Social Change*, p. 86 n. 7.

10. Vowles, "An Inquiry," pp. 8–9. His "Early Evolution of Power Engineering" includes a diagrammatic cross section of a Persian windmill (plate 24) and a photograph of several connected Seistan wind tunnels.

1. Eastern, or Persian, windmills. Sketch from *Windmills*,
by Anne and Scott MacGregor, reproduced by kind permission
of the authors. These ancient mud brick wind tunnels have
little in common with the twelfth-century post-mills.

riod, when they were set to work on the Chinese coast to pump
brine.[11]

The influence of the Persian design on the European windmills is even more dubious. Nevertheless, some authorities have
found it hard to believe that such a complex apparatus could have
been invented more than once, and they therefore have argued for
diffusion of the idea from a single source. This is almost a philosophical analysis, based more on a particular concept of human
nature than on any definite evidence. Equally probable is the
likelihood of independent, perhaps multiple, invention. Some
people believe that this perception testifies more emphatically to
the limitless capacity of thinking beings. It is the position I take.

11. Joseph Needham, *Science and Civilization in China*, 1:245.

The European post-mill seems to be authentically unique. The large, stationary Eastern engines bore hardly any relationship whatsoever to the West's frail wooden contraptions. The European device had complex internal gearing and external quartered blades that spun in a vertical plane. It could be turned completely around to face directly into the shifting northern gusts. In both concept and execution, it appears to have been a totally indigenous invention.

Nonetheless, the mysterious East still captivates people's imaginations; some believe that almost everything appeared there first. The frequently repeated assertion that the crusaders served as intermediaries for the spread of new ideas from East to West is not tenable in this instance, however. What diffusion there was seems to have proceeded in the opposite direction. One twelfth-century writer flatly maintained that Christian soldiers from Germany first introduced the windmill to the Holy Land about 1191.[12]

If the post-mill was a truly European development, which country most probably deserves credit for its perfection? Besides this single reference to German expertise, only three other examples of twelfth-century Continental windmills have been detected: in Normandy about 1180, in Portugal about 1182, and in Provence sometime before 1202. Despite the Dutch windmills' later popularity, no Low Country forebears have been identified before the end of the thirteenth century.[13] Clearly, the numbers of early

12. For more on diffusion, see Chapter 2.

13. Alexander de Lievielle gave the Norman abbey of Saint Sauveur-le-Vicomte a parcel of land near a windmill (*molendino de vento*). This Alexander appears in other sources before 1186. Air and water mills (*molendina tam aure quam aque*) are mentioned in the statutes of Arles (which is near Barbegal, site of the Roman water mill ruins) between 1162 and 1202. These statutes cannot be considered evidence of the existence of any type of mill before 1202. The term "air mill" has not been found in an English context. A charter issued by Count William of Mortain, apparently dated 1105, refers to water mills and windmills (*molendina ad aquam et ad ventum*), but it is clearly a forgery. See Leopold Delisle, "On the Origin of Windmills in Normandy and England," pp.

Continental structures are insignificant compared with the impressive multiplicity of English examples.

The subject of windmills has attracted many distinguished artists and writers, but the search for the earliest prototype seems to have begun in 1851, with publication of perceptive essays by the renowned French scholar Leopold Delisle. Some of his documentation is now questionable and his European examples cannot be dated very precisely, but he advanced the broad thesis of a North Atlantic invention.

Forty-eight years later, in 1899, two indefatigable British writers, Richard Bennett and John Elton, published a multivolume *History of Cornmilling*. Their text focused on Insular practices and featured numerous citations, legendary references, and appropriate manuscript illustrations. It is a treasure trove for anyone curious about the English aspects of this investigation. Interestingly, their synthesis came at the climax of the popularity of water and wind power. Beginning in 1900, electric energy powered more and more facilities.

Early in the twentieth century, English local antiquarians began to notice the increasing disappearance of derelict windmills and made earnest efforts to record and photograph them, usually on a county-by-county basis. One of the most passionate champions of windmills, Rex Wailes, was a practical man who, in addition to appreciating their beauty and history, understood the mechanics of their construction. For many years he fought a lonely battle to preserve the few remaining mills. He gradually gained wide public support and eventually became the driving force of the Wind and Watermill Section of the Society for the Protection of Ancient Buildings.

---

403–6; and his *Etudes sur la condition de la classe agricole et l'état d'agriculture en Normandie au moyen âge*, p. 514. The Arles statutes are reprinted in C. J. B. Giraud, *Essai sur l'histoire du droit français au moyen âge*, 2:208. White discusses the difficulties of assigning twelfth-century dates to these documents in *Medieval Technology and Social Change*, p. 87. For the Portuguese windmill, see Chapter 2, note 13.

Across the ocean, Lynn White, Jr., professor of history at the University of California, Los Angeles, was fascinated by technological progress of all sorts and by its relation to significant human values. He sought to explain the development of windmills in the context of other medieval inventions and of all medieval religious purpose. Because of his style and wit, he popularized the investigation of medieval technology in a way that no one else had.[14]

The fifty-six twelfth-century post-mills identified in this book dramatically strengthen England's claim to be the originator of the Western windmill. They also reveal details about the construction, financing, and public support of the earliest structures. However, the specifics concerning the first post-mill will probably always elude discovery.

The primordial windmills were striking, awkward, almost comical sights. Compared with the massive stone castles and towering Romanesque churches that dotted the landscape, they must have seemed like flimsy toys or tiny sentry boxes on wooden stilts. However, not until a full century after the windmill's invention did medieval artists and writers record their impressions about it. One early literary allusion is distinctly frightening. As he entered the deepest depth of hell, Dante, the immortal Italian poet, perceived a shadowy movement, which he likened to a turning windmill, looming in the distance at eventide.[15]

The first windmills were less frightening than absurd, or whimsical. They were post-mills, little gabled cabins pivoted to rotate atop a strong upright post. On the front were large crossed sails (or arms, or blades, or sweeps, or wings, as they were var-

14. Lynn White, Jr., *Medieval Religion and Technology*, is an important summary of recent research on the development of windmill power. Other significant contributors to our knowledge about windmills include M. I. Batten and Donald Smith, Stanley Frese, Walter Michinton, John Reynolds (whose book has splendid illustrations), John Salmon, Terence Smith, and Hugh Vowles. Their works are cited in the Bibliography.

15. Dante Alighieri, *La Divina Commedia*, ed. C. H. Grandgent and Charles S. Singleton (Cambridge, Mass.: Harvard University Press, 1972), p. 302 (*Inferno*, canto XXXIV, line 6: "par di lungi un molin che 'l vento gira").

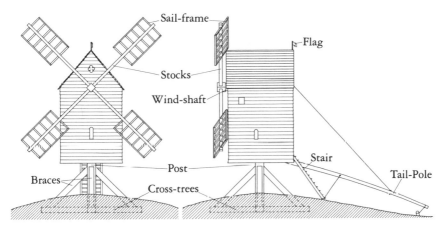

2. Diagram of a typical windmill, drawn by Terence Paul Smith for *History Today* and reprinted with his kind permission.

iously called) consisting of lattice frames covered in board or canvas. Usually four such arms caught the air and spun an attached main drive axle, or wind shaft. Inside the cabin, that horizontal axle was geared to a vertical shaft that rotated the grinding wheels. At first the wind shaft was a level horizontal spindle, but it was later tilted slightly upward for greater stability and efficiency. A sloping tail pole hung from the rear of the cabin. By pushing it, the whole upper structure could be turned on its central post.[16]

The huge upright post that supported the whole building had to be balanced carefully and secured effectively to withstand the terrific vibration of spinning wings and rubbing stones. Therefore, the superstructures of most post-mills were similar, but their foundations varied. Originally, a cut post was probably deeply embedded in the earth and simply braced with diagonal timbers grounded in other holes. The mill supported by this type of post was sometimes called a peg mill, or sunk post-mill.

Technicians quickly learned that the post could be made more

16. John Reynolds, *Windmills and Watermills*, pp. 68–114, includes a perceptive, illustrated discussion of the design and operation of a post-mill.

durable if it were socketed to beams laid crosswise on the ground. These flat oaken supports were called cross trees. Further buttressing could be provided by inclined struts extending from the outer beam ends to the central upright post in an open-ribbed pyramid arrangement. The result resembled the rough wooden bases that are often nailed to modern Christmas trees.

When millwrights desired still more support, they buried the whole lower platform. The added bottom weight was advantageous, but the cross trees and post ends eventually rotted in the damp soil. Thereafter another solution was devised: the lower platform was raised above ground but was fastened upon firmly rooted stone piers that formed a stiltlike tripod.

Although this developmental sequence seems logical, it is difficult to trace the progress of mill experiments and to date the burial and resurrection of the cross trees.[17] Undoubtedly many forms were used simultaneously. It may have taken a considerable time to perfect the pier-based model, for post-mill archaeology and aerial photography indicate a persistent preference for sunken cross trees. The windmill illustrations that finally did appear show the anchored peg mill, the pyramid struts, and the elevated mill on stilts.

The origin of the idea of a revolving structure balanced atop a stationary post is also difficult to discern. Some suggest that the first post may actually have been a tree trunk still rooted in the soil but deprived of its branches. The vivid terminology (cross trees and crown trees) seems to affirm this conclusion. Others claim that the idea derives from the sea, from ship sails swaying about a rigid mast. There are some literary allusions to revolving buildings, but they are hardly convincing evidence.[18]

17. Terence Paul Smith, in "The English Medieval Windmill," discusses post-mill archaeology and suggests a typological sequence for base supports. For information about an early-thirteenth-century post that stood in its own hole and was supported by four inclined struts, but was not connected to any horizontal base, see D. M. Hall, "A Thirteenth-Century Windmill Site: Strixton, Northamptonshire."

18. For more on revolving buildings, see Chapter 2. See also Wilbur Owen Sypherd, *Studies in Chaucer's House of Fame*, pp. 138–55.

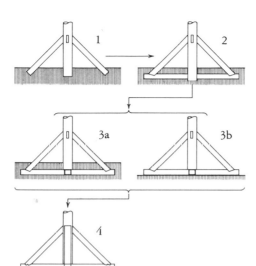

3. Typological sequence
of post-mill supports,
drawn by Terence Paul Smith
for *History Today*
and reprinted with
his kind permission.

The most plausible explanation is that windmill sails were similar to the traditional Roman, or Vitruvian, waterwheels and that an early architect decided to rotate the whole structure as well. Thus, the sails could spin vertically and the cabin could be turned horizontally. In any case, the original design was imaginative and complex. Although easily comprehended thereafter, its execution always required masterly skill.

Pivoting atop the massive main post, but hidden from external view, was the crown tree, a stout transverse beam. Attached to it was the cabin, which held aloft the outside sails and housed the interior grinding machinery. The crown tree and its weighty, awkwardly distributed dependents could be laboriously pushed around in a full circle. (To simulate the effect, try spinning an inverted cup on a pencil or a finger. Note how the cup handle impedes the effort.)

The pivoting cabin, or buck, was small, light, and superbly balanced. Its sides were usually made of clapboard and painted white, or tarred black. Some cabins may have had wicker walls. Early versions had high-pitched gables. In postmedieval models, the roof rafters were curved out a bit, or bowed, to provide more room for the brake wheel and the storage bins. Water mill roofs were sometimes made of thatch, but that material was too heavy for the roof of a post-mill cabin.

The first little cabins must have had limited space. Their internal arrangements can only be conjectured, but they undoubtedly resembled those of later post-mills. On the outside four sails turned the horizontal wind shaft; that axle entered the cabin through the gable just below the roof line. Mounted on the wind shaft was a large brake wheel whose projecting studs meshed with the staves of an upright lantern pinion wheel. The lantern pinion was attached to the vertical power shaft, which turned the millstones. Flour dust was everywhere. With the power drive coming down and the support post thrusting up, there was hardly any room for the grain, the hopper, or the miller.

The long pole protruding from the back of the buck enabled the mill boy to push the whole structure around to greet the shifting head wind. A ladder or set of stairs nailed to the millhouse door swung with the building and gave access to its main floor, high above ground level.

The post-mill was a daring construction. Harmonizing its various parts was an authentic test of the genius of carpenters and millwrights. Surprisingly, such a complex apparatus could be built rather quickly and with only a relatively modest investment of capital—further evidence of craftsmen's skill. One twelfth-century post-mill, for example, seems to have been erected in a single day.[19] More time, expense, and effort were required to establish a water mill. Windmills needed only about one-half an

19. See Chapter 6.

4. An impression of the internal
arrangements of an early
post-mill, from P. S. Jarvis,
*Stability in Windmills*; reproduced
by kind permission of the author.
Note the brake wheel meshing
with the lantern pinion,
thereby turning the millstones.

acre of land, whereas water mills required either a strong, de-
pendable stream or the flooding of several precious acres to create
a mill pond.

The grinding stones used in windmills were identical with
those used in water mills. Unfortunately, they were costly to
purchase and expensive to maintain.[20] There were sources of millstone grit in Derby, Somerset, and Nottingham, but the best
stones came from abroad, usually from the Rhineland, and were

20. Ian Keil, "Building a Post Windmill in 1342."

sold in sets—uppers and lowers. They weighed about a ton apiece and were two to four feet in diameter. Since their radial cutting grooves needed redressing almost every two weeks, prudent millers wanted multiple pairs.

The external characteristics of windmills came to fascinate observers. The sails, buck, and tail pole were all featured in verbal descriptions, manuscript miniatures, wooden carvings, monumental brasses, wall paintings, stone sculptures, and even graffiti casually scratched on church walls. Regrettably, none of the visual representations can be attributed to the twelfth century.

Some graffiti sketches may have been made quite early, but these amateur efforts, which occasionally omitted elements such as the ladder and tail pole, cannot be dated by conventional stylistic or documentary references. Nevertheless, their geographic distribution is suggestive. Two of the churches with graffiti depicting post-mills are in Suffolk near the Abbey of Bury Saint Edmunds, the location of a major dispute concerning windmill franchises (see Chapter 6 and Figure 13). Two other churches with windmill graffiti are close to the ancient post-mill that still stands at Bourn in Cambridgeshire. A graffito in Saint Mary's Church, in King's Walden, Hertfordshire, could be quite old since some parts of the church seem to have been built in the late Norman period. This church is not near a known twelfth-century mill site, but a nearby, still-standing windmill may have been erected atop several predecessors.[21]

21. Saint Mary's Church in Dalham, Suffolk, has two graffiti that show post-mills, one on each respond of the tower; two graffiti depicting tower mills appear on the walls of the belfry. The village still has a windmill. Saint Mary's Church in Lydgate, Suffolk, has several post-mill sketches in various parts of the building; see Violet Pritchard, *English Medieval Graffiti*, pp. 136–37, 149. Saint Mary's Church in Hardwick, Cambridgeshire, has a post-mill drawn on the south doorway, cut through an earlier Latin inscription; ibid., p. 46. The thatched Church of Saint Michael in Long Stanton, Cambridgeshire, has graffiti depicting post-mills on its porch; see Rex Wailes, "The Windmills of Cambridge," p. 97. On Saint Mary's Church in King's Walden, Hertfordshire, see Pritchard, *English Medieval Graffiti*, p. 101. Also see Terence Paul Smith, "Windmill Graffiti at Saint Giles' Church, Tottenhoe."

The earliest datable painting of a post-mill was rendered about 1250–1260, more than a century after the device was invented. It is a playful, four-inch high, blue-and-gold illuminated letter in a manuscript (see the Frontispiece). The professional artist who drew this shimmering scene was English and had numerous other miniature paintings to his credit. His assignment had been to decorate a copy of Aristotle's *Meteorologica*, and he employed the windmill to elaborate the initial letter of the section on wind, thunder, and lightning.

The artist depicted the mill with four blades facing forward and exaggerated the tail pole to achieve his aesthetic purpose. The buck pivoted on a thick post with a weighted base. No ladder was attached, so the mill was evidently quite low to the ground. Since the main post lacked a bottom trestle of sloping quarter boards, the mill probably exemplified the peg, or sunken post, rather than the above-ground type of construction. There were some clever touches, such as the tiny window from which a miller could reef his sails. The painter demonstrated a special sensitivity to surface textures. He realistically rendered the coarse cloth sheets stretched taut across square frames and placed shingles, or wicker strips, on the cabin walls. That cabin appears almost cylindrical rather than rectangular, which was the usual shape.

The miniaturist enlivened the scene by adding a slight, brown-robed boy straining to push round the tail pole and thereby to rotate the whole mill. The lad's strength was further tested by a large sack of flour perched on the tail beam.

The windmill scene was one of three miniatures in this copy of the *Meteorologica*. There were twelve other works of Aristotle and twenty-four more miniatures in the rest of the manuscript. It was quite an expensive collection—undoubtedly a specially commissioned volume.[22]

22. B.L. Harley MS. 3487 fol. 161. For a superb discussion of the manuscript, see Michael Camille, "Illustrations in Hanley MS. 3487 and the Perception of Aristotle's *Libri Naturales* in Thirteenth-Century England."

5. One of the earliest representations of a windmill,
an initial letter, painted in England about 1250–1260,
introduces Book III of Aristotle's *Meteorologica.*
From MS. Ee. 2.31, fol. 130; reproduced by permission of
the Syndics of the Cambridge University Library.
A contemporary of this artist drew a similar, probably
slightly earlier, post-mill (see Frontispiece).

Although the windmill had not previously been a popular
artistic subject, this painter's choice was immediately imitated.
A second English illuminator, also working for wealthy patrons
about 1250–1260, duplicated the first master's design in startling
detail, while retaining his own style. He, too, was decorating a
copy of Aristotle's *Meteorologica.* On the whole, he seems to have
been a more confident artist, less concerned with representing the
brightly patterned background of his subject than with high-
lighting the main activity. This suggests that his was the later
version. Alternatively, the two illuminations may have been mod-

eled on a lost earlier exemplar, or (less likely) an artists' plan book. In either event, the two painters probably knew each other.

The two miniaturists exhibit a striking uniformity of conception, but minor variations can be noted. In the second, less crowded initial, the windmill sails are larger, are viewed from the rear, and are represented laterally rather than diagonally. One sail has disappeared from sight. The cabin seems cylindrical and its walls are depicted as in the first initial. The gable roof is much more prominent and has gained three windows; the millhouse door is now visible, and the flour sack has fallen partly inside. Most interestingly, the central post is clearly supported by aboveground quarter bars. The second mill boy is considerably stronger than his prototype and is dressed in a blue tunic and contemporary snood cap. He seems to be pulling, rather than pushing, the mill round. Both pictures demonstrate that the artists had a pleasant familiarity with their subject.

The sudden appearance of these two illustrations, after such a long period without any post-mill depictions, was not wholly accidental. The *Meteorologica* they decorated, along with the other dozen treatises of Aristotle, constituted a textbook containing all the required reading in natural philosophy for the Oxford University curriculum. In 1250 these works were still new to Western scholars and a standard illustrative program had not yet been established for them. The abstract nature of the material made traditional biblical decorations inappropriate. Thus, the first artist was breaking new ground in preparing his drawings.

The windmill quickly became a very popular motif. One of the most beautiful examples appeared about 1270, but in a psalm book rather than in a scientific textbook. The post-mill, on the second page of the psalter, was part of an elaborate scene depicting the judgment of Solomon (see Chapter 6 and Figure 18). Most painted representations date from the fourteenth century, or later. They were often elements in narrative panels depicting attacks on castles (Figures 16, 17, 19), dissatisfied mill customers (Figures 23, 24), and even Christ being tempted by the devil (Figure 22).

After 1300 windmills began to appear on wooden church furniture, brass memorial plaques, and stone columns. About 1325 a stonemason working on Norwich Cathedral carved a curious scene found in some manuscript illustrations. A customer is depicted bringing grain to a post-mill. She sits astride a horse or ass, but the sack of wheat is atop her head, rather than on the animal's back. Presumably this uncomfortable position was thought to afford the beast some relief from the dead weight of the grain. Wall paintings of post-mills began to appear in the late fourteenth century.

The medieval depictions are remarkably similar to the oldest surviving windmill structures, dating from the seventeenth century. Together they offer a reasonable guide to the earliest design, a highly successful model requiring only minor changes over the centuries.

The major innovation was the stationary tower windmill introduced in the late thirteenth century. This large, picturesque structure, so familiar from representations of Dutch landscapes, rotated only at the top and offered the miller much more interior space and overall stability. Some whitewashed Mediterranean mills (the kind Don Quixote attacked) did not need to rotate at all. Even so, the smaller revolving post-mill retained its popularity and was exported to various parts of the world, including the thirteen British colonies in North America. An eighteenth-century example from Williamsburg, Virginia, was commemorated on a 1980 United States postage stamp that was part of a series honoring the evolution of American windmills from the wooden post-mill to the open steel pedestal.

Several aged post-mills still survive. One of the oldest English examples, the Bourn windmill in Cambridgeshire, dates from at least 1636. Although an impressively large structure, it has the smallest tarred buck of the extant British mills: 10′ 3″ wide, by

6. The Bourn windmill in Cambridgeshire,
dating from at least 1636.
Photograph reproduced by permission of
the *Cambridge Evening News*.

14′ 6″ long, by 31′ 6″ high. The roof has a venerable straight
gable, rather than the more commodious bowed type. The whole
fabrication—arms, cabin, and tail pole—rests on a hoary column
eighteen inches thick. This post-mill has been altered, extended,
and restored several times, but it still preserves the medieval
qualities of economical space, efficient action, and lighthearted
purpose.

The Bourn landmark was badly damaged during a lightning
storm in 1974, a calamity which reflects the paradox that nature
is simultaneously a mill's best friend and worst foe. Fire, rain,

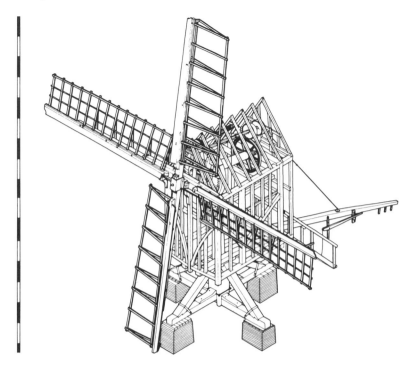

7. A measured drawing of the Bourn windmill,
by John Reynolds, from his marvelous
*Windmills and Watermills*. Reproduced by
permission of Godfrey Cave Associates, Ltd.

wind, and rot can damage any mill. Constant usage and endless
internal vibration are other causes of deterioration. So much de-
pends upon the skill and care of the operator, and on the unpre-
dictability of his environment.

The miller was a rather unusual person in medieval society.
He was a member of the working class, but like the blacksmith,
he was different from most other laborers in that he never farmed.
It was even said that the thumb of a miller's left hand took on a
distinctive shape—a little wider and flatter than the right
thumb—because he was constantly testing the ground meal between his fingers. Millmen probably also endured numerous res-

piratory illnesses because they continually breathed air laden with brown flour dust.

Some twelfth-century millers, operators of windmills and of water mills, were serfs; others were free manorial tenants; still others were prosperous independent freemen. Whatever their precise legal status, most were important community leaders who were frequently called upon to witness transactions in their locality. Charters preserve many of their names but do not reveal much about their lives. Nevertheless, the frequent appearance of millers attests to their growing social and economic importance and suggests the pride they took in their craft. Some even became wealthy enough in their own right to give away their means of livelihood, their mills.

Milling was a desirable occupation, one that was passed quite regularly from father to son. In some localities head millers, or millers-in-chief, supervised several machines simultaneously, both water and wind facilities. Perhaps because of their early success and relative independence, millmen were among the last tradesmen to form guilds for mutual help and professional advancement. Some combined their trade with another occupation, like that of smithing.

In literature the miller, like the archdeacon, eventually became stereotyped as exhibiting greed and coarse behavior. Who can forget Geoffrey Chaucer's witty characterization? The millman was important enough to arouse fear, yet familiar enough to ridicule. Thus he was constantly criticized for sharp practice—for taking as his fee, or multure, more than the one-sixteenth part of ground flour allowed by tradition.

Undoubtedly there were frequent opportunities for such exploitation and profiteering. One such instance and its tragic consequences were graphically illustrated in a fourteenth-century manuscript; the drawing showed a disgruntled female customer burning down the windmill of a man who had cheated her (see Figures 23 and 24). Most early millers, however, seem to have been quite reputable individuals. They certainly possessed unusual physical strength, engineering skill, and commercial acumen. If such qualities sometimes prompted gossip, they also commanded respect.

No special term was used to refer to a windmill miller, but many of the requirements of his craft were substantially different from those of his water mill counterpart. Post-mill work was far more dangerous, and demanded dexterity and sheer muscle power. Like a veritable ship's captain, this miller was forever resetting his engine, putting up sail, stretching canvas, changing the direction of his sheets, and repairing his mast and spars. It is probably no accident that windmill operators developed a complex, colorful language to describe their machines' parts and functions—a lingo far more graphic than that of water millers. On the Continent, windmills, like ships, were given female names. In England, they were given nicknames, usually suggested by their locations.

The successful windmill mechanic needed to develop a sensitivity to the vagaries of local weather and to the eccentricities of his apparatus. He had to be prepared to react instantaneously. A post-mill could easily be bowled over by a tail wind, if a lazy miller had neglected to turn his sails into the gale. Many an operator suffered bodily harm when trying to disengage his gears and reef his sails in a severe storm. His problems multiplied as mills were built ever higher, stretching to catch yet stronger breezes. The timetable of mill work was as unpredictable as the air itself. Countless frustrated millers must have cursed the empty skies while waiting for a zephyr.

Along with the inevitable village church and the occasional baronial castle (some of which would support windmills on their battlements), the post-mill was probably the most prominent structure on the medieval horizon. Indeed, windmills often appear in documents as landmarks to identify nearby acreage. The records emphasize the activities of windmill proprietors, rather than those of mill operators. Accordingly, their ambitions and arguments are highlighted in this study. First, however, a word should be said about the legends that have grown up around early windmills. Then the specifics of the most ancient mills can be tabulated.

# Shadows and Substance

---

*A*lthough labor-saving technology is one of the finest expressions of human ingenuity, the identity of the medieval wizard who first tamed the fleeting European air remains almost as elusive as that of the prehistoric genius who created the wheel. Cumulative research, however, is gradually outlining the range of his activity and reflections of his passage can now be discovered in a variety of sources.

Some reports belong in the realm of fantasy, or even outright deception. A few are speculative, that is, their accuracy is highly probable, but still incapable of verification. Others are clearly based on fact—for example, the details of archaeological excavations in rural areas. The best sources are formalistic legal transactions, but even they have numerous problems of interpretation. Individually, such varied tidbits reveal very little, but collectively they constitute a remarkably coherent mosaic. Therefore, the quest is for the earliest windmills and for whatever details can be elicited about their first decades of life.

Even though all the valid evidence confirms that the post-mill was a twelfth-century invention, some enthusiastic chroniclers and poets have tried to give it greater longevity. One imaginative writer, for example, has claimed that windmills and water mills were both adopted in Wales about A.D. 340. Other commentators, medieval and modern, have given the post-mill a Saxon heritage.[1]

According to one such claim, King Egbert of Kent in 669 gave the Abbey of Saint Augustine in Canterbury part of the Isle of Thanet. Almost eight centuries later, about 1414, the donation was mentioned by Thomas of Elmham, a monk at Saint Augustine's, in his history of the monastery. Thomas illustrated his account with a crude map that included the depiction of a windmill on the slope of the hill at Buckington. The implication was that the gift and the mill existed simultaneously. A windmill did stand there as late as 1799, and it probably had been there since the fifteenth century. Two other post-mills are known to have existed on Thanet in the twelfth century. However, the Saxon origin of Saint Augustine's mill is purely fictitious.[2]

Another attribution to the Saxons is also specious. In 833 Wiglaf, king of Mercia, a rather mediocre monarch who lost and regained his kingdom, allegedly confirmed a long list of the possessions of Crowland Abbey in Lincolnshire. Included among the specific grants was one by a former sheriff named Norman, who had given the monastery two carucates (about 240 acres) of land in Stapleton, as well as two carucates and a windmill in Sutton-next-Bosworth. Both estates are in Leicestershire. Norman was also said to have donated land at Badby in Huntingdonshire. Wiglaf's lengthy charter is contained in a history of Crowland ascribed to Abbot Ingulf (d. 1109).

---

1. For a summary of the fanciful theories concerning the origins of windmills, see Bennett and Elton, *History of Cornmilling*, 2:224–34. The claim for Wales was made in the so-called *Gwentian Brut* (or *Brut y Tywysogion*), an 1801 forgery by Iolo Morganwg (Edward Williams, 1747–1826) that was not exposed until 1928; see John E. Lloyd, "The Welsh Chroniclers," pp. 376–77.
2. Bennett and Elton, *History of Cornmilling*, 2:248.

Everything about this peculiar chronicle and about the list is suspect. Scholars have not been able to determine exactly when, but the history may have been written as many as four centuries after the death of the abbot. Parts of it are obviously fanciful. Whether the author incorporated earlier charters into the chronicle or whether he made up most of them cannot be determined.

The 833 charter credited to Wiglaf is clearly a forgery. A few of his deeds do exist and are known to be genuine, but modern scholars universally reject this particular text's authenticity. Moreover, the charter's pretentious term for windmill, *molendinum ventricum*, was not used until the end of the twelfth century, and that, in itself, is sufficient to render the whole document suspicious.[3]

Nevertheless, a forgery can contain a trace of truth. This fabricator was not primarily interested in windmills. Rather, by making grandiose royal declarations he intended to enlarge, or possibly merely to reconfirm, his monastery's total endowment. Domesday Book indicates that Crowland Abbey did, in fact, own manors at Sutton and Stapleton before the Norman Conquest.

Moreover, Sheriff Norman can be connected directly with Sutton. An unusually splendid scroll, dating from the late twelfth or early thirteenth century, is illustrated with scenes from the life of Saint Guthlac (d. 714), the hermit patron of Crowland. Its last panel depicts a crowd of thirteen benefactors pressing against the saint's altar with their various gifts. The group is led by King Ethelbald, founder of the abbey, and includes Sheriff Norman,

---

3. The charter has been printed a number of times; see William Dugdale, *Monasticon Anglicanum*, 2:109; Walter de Gray Birch, *Cartularium Saxonicum: A Collection of Charters Relating to Anglo-Saxon History*, A.D. 430–975, 4:567, charter 409; John M. Kemble, ed., *Codex diplomaticus Aevi Saxonici*, 1:301, charter 233. See also Peter H. Sawyer, *Anglo-Saxon Charters: An Annotated List and Bibliography*, charter 189. Some versions attribute the windmill grant to King Beorred, Wiglaf's successor; see John Nichols et al., *Bibliotheca Topographica Britannica*, vol. 3, appendix, p. 15. On the forgery of the collection, see Henry Thomas Riley, "The History and Charters of Ingulfus Considered"; and W. G. Searle, *Ingulf and the Historia Croylandensis*, pp. 167–69.

who is stating that he donated land in Sutton and Stapleton. However, Norman was not alive in 833, the date of the Wiglaf charter. He lived in the early eleventh century and was a brother of Earl Leofric of Mercia, husband of the famed Lady Godiva. Norman was murdered by King Canute (1016–1035).[4]

Although these historical associations constitute only a rope of sand, they do suggest a tentative explanation. Certainly there was no Saxon post-mill in Sutton in 833, and none in 1050. However, a windmill did exist in Sutton when the pseudo-Ingulf concocted his tale. Memory of it may have been so ancient that later generations thought that it predated the Norman Conquest and, logically, linked it with the manor's oldest known owner, Norman the sheriff.

Of course, from the viewpoint of the Crowland monks, it was desirable that their properties have an ancient royal confirmation. Hence the gratuitous graft to King Wiglaf. The sheriff and his mill were merely items in a long list of benefactions, true and false, that the forger wanted confirmed. However, King Wiglaf and his Leicestershire lands should not be dismissed from memory too quickly, for as we will see, they will again be linked with the earliest windmill.

Another sheriff also had a dubious connection with early windmills. Picot was one of the many rough warriors who made his fortune as a result of the Norman Conquest. He gradually acquired land in several counties and later served as sheriff of Cambridgeshire. Only an eighteenth-century antiquarian linked him to a post-mill. Without offering any evidence, John Bowtell asserted that Picot built a windmill in Cambridge.[5]

In truth, Picot was involved in mill construction. Domesday Book records that he built three mills in the town of Cambridge alone. To erect them, Picot laid waste pastures and destroyed

---

4. George Warner, ed., *The Guthlac Roll*, frontispiece. For a discussion of Sheriff Norman, see Frances S. Page, *The Estates of Croyland Abbey*, pp. 5–7.

5. H. P. Stokes, "The Old Mills of Cambridge," p. 223.

8. An overshot water mill, drawn about 1220–1230. The great
wheel is depicted with attached axle and internal gears
(a pit wheel and a lantern pinion). Similar gears would
have been used in early windmills. The accompanying text,
a copy of Saint Anselm's *De Similitudinibus*, compares the
mill to a human heart. From Cotton MS. Cleopatra C XI,
fol. 10; reproduced by permission of the British Library.

several existing houses. Such high-handed destruction of peasants'
property by the conquering Normans was not unusual, but Picot
also tore down one mill belonging to the monks of Ely Abbey
and another belonging to Earl Alan of Brittany. The effort paid
off handsomely. In 1086 Picot's mills yielded nine silver pounds,
a quite extraordinary profit. Some financial gain must have come
from the sale of fish caught in the mill waters of the River Cam.
Whether any part of it was attributable to a windmill is a yet
unanswered question.[6]

There is no convincing evidence that post-mills existed in the
eleventh century. Such a possibility is vaguely suggested by the
use of the Old English word *Millichope* (apparently meaning "the
valley by the mill hill") in the Domesday survey of 1086 to refer
to a hamlet in Shropshire. Domesday Book also occasionally listed

6. On Picot, see *Liber Eliensis*, pp. 205–7, 210–12; Edward Miller, *The Abbey and
Bishopric of Ely*, chap. 1 passim.

a "winter mill." This has been interpreted to mean a water mill activated by seasonal streams. An alternative definition might be a wind-powered mill that was used when water froze. It is also interesting that Picot and his wife exhibited personality traits similar to those of verifiable post-mill advocates. The couple was generous to religious houses, friendly with physicians, and absolutely determined to eliminate competition. On the other hand, Archbishop Anselm of Canterbury (1093–1109), a very well informed thinker, did not mention any wind-power features in his extended moral comparison of the mill with the human heart.[7]

A few relevant coincidences are intriguing. The principal seat of Picot's barony was Bourn, just a few miles outside Cambridge. His successors in that fief included the younger half brother of Walter Fitz Robert, one of England's most aggressive windmill champions. The 350-year-old post-mill still standing at Bourn may have had several predecessors. Two nearby churches have windmill graffiti. One of Sheriff Norman's outlaw descendants was active in the area following the Norman Conquest.[8] Someday these coincidences may be strengthened by a solid archaeological or documentary discovery that will give Bourn enduring significance in the chronicle of early windmill development.

The victory at Hastings Field in 1066 adds one other, wholly fictitious detail to the story. An anonymous poet of Shakespeare's day (some identify him as the great bard himself) composed an undistinguished comedy called *Fair Em*. The heroine was a Saxon whose father, Thomas Goddard, had been deprived of his noble position by the conquerors and was forced to support himself by

7. For the *Domesday Book* reference (fol. 258v) I am grateful to Dr. Robin Fleming, who is preparing a computerized data base of the 1086 survey. Also see Eilert Ekwall, *The Concise Oxford Dictionary of English Place-Names*, p. 326. On Saint Anselm's treatise, see Figure 8 and R. W. Southern and F. S. Schmitt, eds., *Memorials of Saint Anselm*, pp. 53–55, 305.

8. Ivor J. Sanders, *English Baronies: A Study of Their Origin and Descent, 1086–1327*, p. 19. Saher de Quency II (d. 1190) was Walter's half brother. Norman's descendant was Morcar of Northumbria.

operating a windmill in Manchester. Em delivered him from such a "base" trade by marrying one of the Normans.[9] The play demonstrates that an Elizabethan author thought the post-mill was at least 500 years old.

Two literary works—one dealing with Charlemagne and one with Andalusia—made more significant contributions to windmill history. Sometime between 1050 and 1250, an anonymous Anglo-Norman poet wrote *The Journey* (or the *Pilgrimage*) *of Charlemagne to Jerusalem and Constantinople*. Although the poem was probably composed early in the twelfth century, the only known manuscript was copied in England about one hundred years later.[10]

Whether this humorous epic was written in Britain or on the Continent, and whether it is a unified work or the union of two disparate parts, are unanswered questions. There is a dispute concerning whether it was a heroic chanson sung for chivalrous lords

---

9. *Fair Em*, in the Tudor Facsimile Texts.

10. The literature about this poem is voluminous, but not all of it is easily accessible. The original, from the British Library, King's Library MS. 16 E VIII, is now said to be lost. It was published in 1836 as *Charlemagne, an Anglo-Norman Poem of the Twelfth Century*, edited by Francisque Michel; and in 1913 as *Karls des Grossen Reise nach Jerusalem und Constantinopel*, edited by Eduard Koschwitz (this is only one of several editions). For a brief commentary on the poem, see Laura Hibbard Loomis, *Adventures in the Middle Ages*, esp. pp. 70–75. See also the studies by Jules Horrent, including "La chanson du pèlerinage de Charlemagne et la réalité historique contemporaine," which shows that the date of the work cannot be precisely determined. About 1400, a Scandinavian version of the poem was incorporated into the *Karlamagnus Saga*, edited by Constance Hieatt, esp. 3:190. Margaret Schlauch, in "The Palace of Hugon de Constantinople," compares the poetic turning palace to what is known of Byzantine palaces. Alfred Adler, in "The *Pèlerinage de Charlemagne* in New Light on St. Denis," contends that the poem was written shortly after the Second Crusade (1147–1149). I am grateful to Professor Ellen Kosmer of Worcester State College for informing me of this spinning palace. A *chanson de geste* (song of heroic deeds) entitled *Lorrains*, or *Girart de Roussillon*, was written by a contemporary of King Philip Augustus (1180–1223). Included were realistic descriptions of military horrors, such as, "In the cities, in the towns, and on the small farms windmills no longer turn, chimneys no longer smoke, the cocks have ceased their crowing, and the dogs their barking"; quoted by Achille Luchaire in *Social France at the Time of Philip Augustus*, p. 261. There is no way to date the windmills in this plaintive passage, but its author considered them a common sight.

and ladies, or whether its bawdy passages were intended for the supposedly coarser townsfolk.

The plot is simple. Charlemagne foolishly asks his court if there is any ruler greater than himself. His wife imprudently replies that some say the emperor at Constantinople is his match. Angrily, Charles shuts her up in a castle. Then he and his companions go off to investigate the claim, piously visiting Jerusalem along the way. This is pure fiction, of course. The lord of the Franks never traveled to the East, although this assertion is also found in other sources. Nonetheless, the poet's description of the outbound cavalcade is rather realistic. After the pilgrims leave the Holy Land, they see many wondrous sights, including the Byzantine emperor tilling the soil with a huge golden plow.

This majestic individual invites the strangers to his palace, which is truly marvelous to behold—a luminous circular vaulted building set on a hollow column. It is adorned with statues that sound trumpets and broadcast music when a breeze comes up. As the awestruck Franks enter, a sharp wind blows in from the sea and spins the whole building around, knocking them to the ground. The poet says the palace turned "just like the shaft of a mill." The embarrassed knights stumble back to their feet and, though somewhat shaken, accept an invitation to spend the evening.

In order to bolster their courage, they drunkenly brag about feats they could perform themselves. One would hurl a heavy golden bowl an extraordinary distance. Archbishop Turpin would single-handedly divert the course of the river. And the usually level-headed Oliver displays a surprisingly lusty streak, maintaining that he would have intercourse with the lovely Byzantine princess 100 times in a single night.

A spy hidden in the hollow pillar overhears them and reports their boasts to his master, the emperor, who the next day furiously challenges the Franks to prove their claims. Unfortunately for him, they are able to fulfill them one way or another, and he is forced to become Charlemagne's vassal. Victorious and assured of their lord's prowess, the Westerners gallop back home.

How different is the tone of this epic from that of the contemporary *Chanson de Roland*. Yet, how fitting it is that the mightiest hero of Christendom, the fabled Charlemagne, should wend his way into the bypaths of wind-power technology. Fantastic spinning houses also appear in other literary works, especially those with Celtic origins.[11] However, this Byzantine palace was turned by the wind, and its action was compared to that of a mill shaft. One immediately thinks of the elevated rotating buck of a post-mill and its spinning arms. Indeed, some would translate *com arbre de molin* (line 372) as the sail of a mill.[12] But a case can be made that the simile pertained to any mill shaft. The ambiguity is compounded by use of the unspecific term "mill," for both water mills and the earliest windmills were often referred to simply as mills.

The sense of the whole passage suggests that the poet had a windmill in mind, however. If this could be proved, and if this work could be precisely dated and its author identified, the poem would be powerful evidence serving to localize the development of the post-mill. Without such information, the verse remains a fascinating reference demonstrating that rotary buildings were familiar to Anglo-Normans, if only in their imaginations. It is also worth noting that the turning palace is integral to the whole second half of the tale; it is not a later insertion. The idea of a wind-rotated building is thus as old as the poem and is depicted with whimsy and wonder, not with the fear and dread found in some Continental windmill descriptions. The manuscript is English in origin, and the poem seems addressed to prosperous, fun-loving types; it is middle-brow literature.

A journey to the East and a fictitious observation of applied

11. See Sypherd, *Studies in Chaucer's House of Fame*, pp. 144 ff.

12. See Hieatt, *Karlamagnus Saga*, 3:190. "Wing of a mill" would be another translation.

wind power raise one of the most persistent questions about wind-mill origins. Was the machine an independent European invention? Or was it the result of a diffusion of ideas and the imitation of exemplars from the Orient? More specifically, can the wooden post-mill be traced to the mud brick tunnels of Seistan? Another literary work unconsciously addresses this issue.

About the time of the Norman Conquest, Abu Zayd Ibn Mukana Alisbuni Alqabdaqi (Abu Zaide Ibne Mucane Alisbuni Alcabdaque), a romantic Andalusian poet from Alcabideche, near present-day Lisbon, nostalgically sang the praises of his hometown. In part he declaimed, "If you are a man of decision, you need a mill that works with the clouds, without depending on streams."

Ibn Mukana Alisbuni was a countryman who went to court and wrote bucolic verse. He had a genuine feel for the land and extolled crops, wine, love, sports, and famous rulers. Frequently he was quite specific in his praise. Later in this particular poem, for example, he refers to the large grain crop produced in Alcabideche in a good year. Yet his language provides only ambiguous clues to prevailing technology. Although he mentions neither air nor wind, he seems to be alluding to some type of wind-powered mill. But which kind? The whitewashed round towers with asterisk-spoked arms that eventually dotted the whole Mediterranean basin? The rickety revolving post-mill of Western Europe? Or the sturdy stone tunnels of Baluchistan?

Although it is impossible to verify such a transmission, news of the Seistan wind machine could have made its way to Islamic Iberia. The chronology seems approximately correct. However, thirteenth-century cosmopolitan Arabic writers in the Levant seemed so surprised upon learning about the post-mill that it is difficult to believe that an eleventh-century poet from Alcabideche could have known about the apparatus. Even more to the point, the earliest windmill to appear in unambiguous Iberian records is dated 1182—well after the Christians had reconquered much of the Iberian Peninsula. Interestingly, it was reported to have been erected near Lisbon.

Commercial, military, religious, and scientific contacts between England and Iberia were well established by the twelfth century. Lisbon was even captured by English crusaders in 1147. Knowledge of the Western-style post-mill could therefore have traveled just as easily from England to Portugal, as it could in the opposite direction. That the Portuguese contributed to windmill development is a tantalizing, but unsubstantiated, conclusion.[13]

The idea of a wind-powered mill could also have been diffused westward along other routes. Certainly, there were multiple contacts between East and West. Diplomats, merchants, and pilgrims had been clogging the narrow inland roads and broad sea-lanes for centuries. Thousands of European soldiers and their attendants had been marching to and from Constantinople and Jerusalem ever since the First Crusade (1096–1099). Any of these travelers could have been a plausible intermediary.

Military orders, such as the Knights Templar and the Knights of Saint John of Jerusalem (the Hospitallers), operated a network of charitable institutions throughout Europe and the Near East.

13. Professor Ellen Kosmer called my attention to an entry in the *Michelin Green Guide to Portugal* (1st English ed.; London: Michelin Tyre Co., 1972), p. 56. It mentions a monument to Ibne-Mucane at Alcabideche, reports that he is believed to have been the first European poet to sing the praises of windmills, and quotes him: "If you are really a man, you need a mill which turns with the clouds and is not dependent on water courses." Professor A. H. de Oliveira Marques of the University of Lisbon kindly supplied further information about this Andalusian poet. A Portuguese translation of Ibn Mukana Alisbuni's poem is included in Antonia Borges Coelho, *Portugal na Espanha Arabe*, 4:335–36. A French version is found in Henri Pérès, *La poésie Andalouse en Arabe classique*, pp. 200–201; a few details of his life and other excerpts are given on pp. 57, 97–98, 239, 345, 372, 375, 410, and 420. Professor Marques's own important study, *Introdução a história da agricultura em Portugal*, includes details about the 1182 windmill (see p. 195). Maurice Daumas, in *A History of Technology and Invention*, 1:110–11, suggests that early windmills could be found as far west as Portugal but gives no evidence or date. Bertrand Gill, in the same volume, says the windmill was probably an importation from the Arabs (p. 429) and suggests that "in Spain, windmills may have been turning in the region of Tarragona since the tenth century" (p. 459), but again, no evidence is offered.

Their members were almost professional travelers and hundreds of Englishmen joined their ranks. It is possible that one of them saw wind machines in the Levant and carried the news back to Europe.

Tempting as this conclusion is, not one shred of evidence supports it. Indeed, persuasive arguments invalidate it. Crusaders are known to have transmitted knowledge about windmills eastward, rather than westward. The First and Second Crusades provide no pertinent details for our study, but the Third Crusade (1189–1192) offers a few unusual sidelights. Its English contingent included many windmill enthusiasts. King Richard, or his late father Henry II, had granted a windmill in Dunwich to the Knights Templar. Hubert Walter, the bishop of Salisbury, served heroically in Palestine and later gave the tithes of two windmills to a Canterbury hospital. William de Forz, count of Aumale (1190–1195), granted a Yorkshire post-mill to Meaux Abbey just before he left for Jerusalem. Earl Robert of Leicester and his steward, Arnold Wood, another Crusade hero, both supported post-mills in Leicestershire, as did Stephen of Camais in Norfolk. Some of these men were also related to windmill proprietors.

The Englishmen who joined this Crusade sailed to the Holy Land, rather than following the traditional long overland route. Some of their battles were sea encounters. According to one late report, during the siege of the port of Acre in July 1191, King Richard had mills with sails built on the decks of his barges and galleys. The terrible noise of their grinding Rhine stones was said to have frightened the Saracens.[14] Shades of Dante's *Inferno* and Don Quixote's fearsome monsters! This account hardly seems credible; probably its author confused a windmill with some sort of catapult. It does, however, lend credence to the assumption that the Moslems would not have known what a windmill was.

---

14. Peter Langtoft, *Chronicle in French Verse*, 2:79–81. Bennett and Elton, *History of Cornmilling*, 2:233, includes a florid translation of the poem, according to which the sails were black and blue, green and red. Langtoft died about 1307.

That belief is also strengthened by an even more intriguing contemporary report. During the Third Crusade, two young idealists, a Norman named Ambrose and an Englishman named Richard, decided to share one another's company and to keep a journal of their experiences. Soon after 1192 each wrote about the adventure, Ambrose in French and Richard in Latin. The resulting books are in close agreement on most incidents, but there are some curious differences. According to a minor interpolation in an early copy of Ambrose's work, the Germans made a windmill (*molin à vent*), the first of its kind in Syria. Oddly, the parallel passage in Richard's Latin account mentions a donkey mill (*mola assinaria*). Yet that could hardly have been a new idea to the Easterners. Regardless of such inconsistencies, the sense is clear: the post-mill was thought to have come from west to east.[15]

Ambrose's attribution of the post-mill to the Germans is somewhat puzzling because there are no indications that Germany had windmills in the twelfth century. Perhaps he meant Anglo-Saxons rather than Germans.

That the wind device was unknown in the Islamic world is further supported by a comment made in 1206 by an Arab writer in Edessa, which had been under crusader control some years earlier. He claimed that the notion of mills driven by the wind was nonsense, for the wind was too fickle to power such a machine. Evidently he had heard about post-mills but had never seen one.[16]

The unique Western pedestal design, the multitude of English twelfth-century post-mills, the universal scarcity of contemporary references to the existence of windmills elsewhere, and the unself-conscious sophistication of English descriptions constitute a strong quadruple argument that the new technology was devel-

15. Kate Norgate, "The *Itinerarium Peregrinorum* and the *Song of Ambrose*," pp. 523–47. See also White, *Medieval Technology and Social Change*, p. 87; White, *Medieval Religion and Technology*, p. 238; and Mary G. Cheney, "The Decretal of Pope Celestine III on Tithes of Windmills, JL 17620," p. 64 n. 3.

16. White, *Medieval Religion and Technology*, p. 223.

oped by the English, and certainly by the Europeans, rather than diffused from the Orient, through either Portugal or Palestine. It is now time to demonstrate that English inventiveness directly, by discussing the linguistic, documentary, and archaeological evidence and by listing all the identifiable English twelfth-century post-mills.

Medieval writers often failed to distinguish the type of mill they were recording. The common Latin term *molendinum* meant any mill device, mill house, or milling place. Mill devices included hand mills and those activated by animal tread, but in early English usage the reference was usually to a water mill. Yet in the absence of an allusion to a mill pond, a sluice, or an aqueduct, there is always the possibility that a windmill was meant. The scribes at Chichester Cathedral and at Hickling Priory, to cite but two examples, regularly used the vague word *molendinum*, but other records show that they meant a windmill. A lack of specific terminology is also found in the field of medicine. More than half of the Anglo-Norman medical practitioners sometimes casually identified themselves only as masters, not physicians.

When a precise identification did begin to appear, it was not a new word but an extension of the basic term. The Latin language reflects how gradually the new technology was incorporated into everyday life. The oldest Latin term used for windmills was evidently *molendinum ad ventum*. It was the most popular term in the twelfth century, for it appears in references to half of the identifiable windmills. I am employing the modern terms post-mill and windmill interchangeably, but strictly speaking, only the latter is a correct translation. Anglo-Norman people had no reason to distinguish among the types of windmills, since the stationary tower mills had not yet been developed.

There were many early variations in windmill terminology, all consisting of the basic noun followed by a phrase or a single word. Besides the popular *molendinum ad ventum*, contemporaries often used *molendinum de vento*, *molendinum venti*, and *molendinum venticum.* Occasionally, they also used *molendinum ventile, molendinum venteritium,* and *molendinum ventricum.* This last, more elegant, phrase became very prevalent in the thirteenth century, when *molendinum per ventum, molendinum ventosum,* and *molendinum ab vento* also appeared.

The rather grandiose word *ventorium* was employed at least once in an early agricultural context, but linguists believe this was a reference to a winnowing fan, not to a wind-powered engine.[17] The French used the phrase *molendinum aurum,* or air mill, once before 1202, but this term has not been found in Insular sources.[18] The term for a water mill also began to change. For the most part, the simple *molendinum* was usually employed, but some scribes used *molendinum ad aquam* and, later, *molendinum aquaticum.*

Many scribes used *molendinum* to mean both a windmill and a water mill. Indeed, half of the identifiable windmills were at one time simply called mills. The terms "mill" and "windmill" were frequently used in the same document to refer to the same installation, as they are today. One hundred and fifty years after the invention of the post-mill, people were still using *molendinum* and *molendinum ad ventum* interchangeably.[19]

Most early references to English post-mills appear in laconic records—usually undated real-estate deeds. Thousands of medieval charters and writs have weathered the centuries, but only four

17. About 1150, Ardleigh manor in Hertfordshire was granted to Master Albert. Its possessions included one hand mill, two winnowing fans (*vannii*) and a ventorium, defined as a winnowing fan in R. A. Latham's *Revised Medieval Latin Word List from British and Irish Sources,* p. 507. Albert and his prebendary successors, Nicholas de Sigillo (d. 1187) and Master Henry of Northampton, were highly educated men who might have used a fancy term for a post-mill. Certainly there was a windmill in Ardleigh before 1222. See Guildhall Museum, Saint Paul's Cathedral Muniments, MS. W. D. 4 (Liber L), fols. 45v–49; *Domesday of St. Paul's,* pp. 21, 137; Marion Gibbs, ed., *Early Charters of the Cathedral Church of St. Paul's, London,* p. 59, charter 82.

18. See Chapter 1, footnote 13.

19. For example, the scribe who composed a 1263 charter about a windmill belonging to William Fitz Ralph; B.L. Harley MS. 742 (Spalding Priory Cartulary), fol. 10.

of the original documents pertaining to windmills have survived. Fortunately, the contents of some of the others were preserved in cartularies, or charter registers. For the sake of convenience and security, landowners, especially large ecclesiastical corporations, had copies of all their documents included in such collections, often decades after the transactions had been completed.

Inhabitants of the same locality knew quite a bit about each other in Anglo-Norman times, and public memory was the traditional custodian of information concerning official business. Enforceable contracts were therefore invariably oral communications, frequently accompanied by ceremonial formalities. For example, many grants to churches were symbolized by placing a knife or staff upon the altar. Charters merely confirmed verbal commitments. Since participants and observers were familiar with the relevant issues, the few written agreements, confirmations, donations, and proclamations that do survive have a no-nonsense air. Descriptive details about windmill age, design, and use thereby escaped report, for they were assumed to be common knowledge.

In addition, many cartulary scribes omitted the names of charter witnesses. If donors, recipients, or attestators were recorded, the chronology of an undated document can sometimes be estimated by ascertaining key dates in the careers of, and the dates of death of, such individuals. Although such a determination indicates a terminus before which a mentioned post-mill necessarily existed, it cannot specify when in a benefactor's or signatory's lifetime the structure was created.

To complicate matters further, twelfth-century parents selected monotonously few Christian names for their children, and even fewer surnames were then common, so it is frequently difficult to decide which individual was involved with a particular document. For example, the Cleydon windmill in Oxfordshire was built by a Henry d'Oilli, but three generations of males in his family bore that name in the twelfth century. It was a rare machine that was as well identified as the windmill at Cardington in Bedfordshire, which was described as "newly constructed" in a writ clearly dated 1260.[20]

Archaeology may soon allow researchers to answer some of these chronological questions, and the result may be an increase in the windmill total. Until about the mid-1960s, windmill mounds were frequently mistaken for prehistoric round barrows, or even for early medieval castle mottes, especially since millers often dug a ditch around mill perimeters to obtain extra dirt to heighten a mound, or to erect a barrier circuit in order to keep cattle away from the revolving sails. Of course, an ancient barrow or an abandoned tower hill also occasionally served as a windmill platform.

Archaeologists are just beginning to become interested in windmills, and their future excavations will probably produce many additional details. Even without digging, perceptive field walkers could easily multiply the number of identifiable post-mills just by demonstrating the absence of abundant water at certain documented mill sites. Post-mill mounds that have been plowed over and leveled can frequently be detected through aerial photography as crop marks in the soil. These linear shadows reveal buried field structures, such as circular trenches with internal bisecting depressions—remnants of the cross trees' foundations. Some sites, such as that of the two abandoned post-mills at Kirkby Bellars in Leicestershire, are not mentioned in available records.

Several of the oldest windmills were built in the lonely open fields; others were erected next to older water mills, often near bridges at river crossings. Their location suggests that practical millers may have first used their new machines as alternative sources of power, perhaps for times when water levels dropped in stagnant ponds and shallow streams, or when the great paddle

---

20. *Newnham Cartulary*, 1:68, charter 109.

9. Aerial view of unexcavated windmill mounds at Kirkby Bellars,
Leicestershire. Note the distinctive depressions in the tops
of the two mounds caused by the decay of the cross tree beams.
Photographed by J. K. S. St. Joseph in May 1966, looking north.
Reproduced by permission of the Committee for Aerial
Photography, Cambridge University Collections, copyright reserved.

wheels became clogged with ice. Perhaps the idea of a wind-driven mill sprang into someone's mind when he observed a dry water wheel spinning crazily in the gusty wind.

The oldest known excavated windmill site is an isolated earthwork just east of the village of Great Linford in Buckinghamshire. This low mound, once grandly called Windmill Hill, had a C-shaped ditch around it and was right beside the medieval path to Newport Pagnell. Nothing is visible on the mound surface now, but the ditch could have served as a track for a long tail pole, particularly if the pole were attached to a cartwheel for easier pushing.

Excavation of this site in 1977 revealed a set of substantial oaken supports (that is, rather typical sunken cross trees) packed in limestone. No other datable artifacts were uncovered, but when the timbers were subjected to radiocarbon analysis, they yielded a calibrated date of $1220 \pm 80$ years. Thus, this post-mill could easily have had a twelfth-century origin. However, its initial documentary citation was a notice that in 1303 it belonged to a William le Waleys.[21]

Sophisticated determination of tree ring sequences may someday enable British archaeologists to estimate the dates of such large wooden artifacts quite accurately. Ancient Indian timbers are routinely dated in this manner in the southwestern United States today. Clay was often used as packing around buried cross trees, presumably as a preservative to protect lumber from the air and dampness.

Searching through old documents has revealed more post-mills than has digging in the dirt. Table 1 is a register of the locations, first datable appearances, and sponsors' names of the

---

21. R. J. Zeepvat, "Post Mills and Archaeology." For reports of other recent excavations, see the bibliographic entries under J. R. Earnshaw, D. M. Hall, P. S. Jarvis, S. V. Pearce, M. Posnansky, and T. P. Smith. For an informative discussion of early post-mill design, see P. S. Jarvis, *Stability in Windmills and the Sunk Post Mill*, which is based on archaeological, artistic, and mechanical evidence.

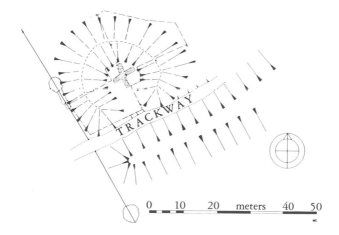

TRACKWAY

0   10   20   meters   40   50

10. The post-mill at Great Linford, 1977. A photograph
showing the remains of timber cross trees and surrounding
limestone packing and a plan of the excavation site.
Reproduced by courtesy of R. J. Zeepvat and the Milton Keynes
Archaeological Unit, Buckinghamshire County Council.

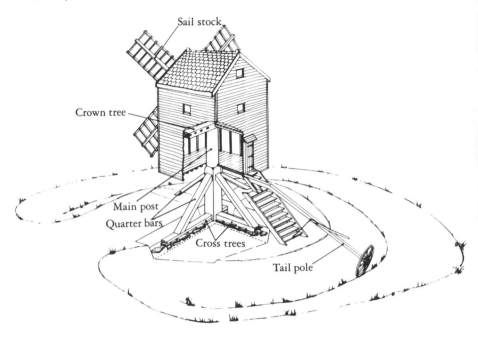

11. Reconstruction of the Great Linford post-mill,
showing main structural components. The artist and
site supervisor, Robert J. Zeepvat, generously amended
his original drawing, published in *Current Archaeology*, to reflect
recent interpretations. Reproduced by kind permission of
R. J. Zeepvat and the Milton Keynes Archaeological Unit,
Buckinghamshire County Council.

earliest known post-mills. Since many mills were given away as
gifts, the intended beneficiaries are also listed. (To facilitate reference, the specific details and documentation for each windmill
have been placed in the Appendix, where the entries are arranged
in alphabetical order by county. Inclusion of this gazetteer con-
veniently eliminates the need for many citations that would otherwise have been included in the footnotes.) It should be remembered that the true number of twelfth-century English windmills
has probably been grossly underestimated.

# Table 1
## TWELFTH-CENTURY ENGLISH WINDMILLS

| LOCATION | DATE | PATRON | BENEFICIARY |
|---|---|---|---|
| **BEDFORDSHIRE** | | | |
| 1. Renhold | −1166 | Gilbert Avenel | Newnham Priory |
| **BUCKINGHAMSHIRE** | | | |
| 2. Dinton | c. 1184 | Agnes of Mountchesney | Godstow Abbey |
| 3. Evershaw | −1200 | Hugh of Evershaw | |
| **CAMBRIDGESHIRE** | | | |
| 4. N—— | −1191 × 1196 | M——, a knight | |
| 5. Silverley | −1166 × 1194 | Reginald Arsic | Hatfield Regis Priory |
| **ESSEX** | | | |
| 6. Henham | −1200 | William Fitz John | |
| **KENT** | | | |
| 7. Canterbury | c. 1200 | Prioress Juliana | Eastbridge Hospital |
| 8. Canterbury | c. 1200 | Stephen Fitz John | St. Augustine's Abbey |
| 9. Chislet | c. 1200 | Abbot Roger | St. Augustine's Abbey |
| 10. Monkton | −1198 | William Wade | Christ Church Priory |
| 11. Reculver | −1195 | Archbishop Hubert Walter | Eastbridge Hospital |
| 12. Romney | −1190 | Adam of Charing | Romney Hospital |
| 13. Westhallimot | −1195 | Archbishop Hubert Walter | Eastbridge Hospital |
| **LEICESTERSHIRE** | | | |
| 14. Wigston Magna | −1169 | Simon Wykingeston | |
| 15. Wigston Parva | −1137 | William the Almoner | Reading Abbey |

*Continued on next page*

| LOCATION | DATE | PATRON | BENEFICIARY |
|---|---|---|---|
| **LINCOLNSHIRE** | | | |
| 16. Friskney | −1189 × 1194 | Gilbert of Benniworth | Ormesby Priory |
| 17. Hogsthorpe | −1200 | Beatrice of Mumby | Barlings Abbey |
| 18. Manby | −1200 | Walter Fitz Alan | Grimsby Abbey |
| 19. North Elkington | −1189 × 1194 | Gilbert of Benniworth | Ormesby Priory |
| 20. Swineshead | −1168 | William of Amundeville | Swineshead Abbey |
| **MIDDLESEX** | | | |
| 21. Clerkenwell | c. 1144 | Jordan Bricett | Clerkenwell Nunnery |
| **NORFOLK** | | | |
| 22. Attelborough | −1198 | Ralph of Burg | |
| 23. Betwick | c. 1200 | Hugh de Albini | Wymondham Priory |
| 24. Burlingham | −1200 | Jocelin of Burlingham | Hickling Priory |
| 25. Fincham | −1200 | Philip and Emma Burnham | Castle Acre Priory |
| 26. Flockthorpe | −1198 | Stephen of Camais | Wymondham Priory |
| 27. Hempnall | 1136–1198 | Walter Fitz Robert | Bury St. Edmunds Abbey |
| 28. Herringby | c. 1200 | Walter of Herringby | Hickling Priory |
| 29. Hickling | c. 1185 | Theobald of Valeins | Hickling Priory |
| 30. Palling | −1200 | Bartholomew of Beighton | Hickling Priory |
| 31. Rollesby | c. 1200 | Beatrice of Somerton | Hickling Priory |
| 32. Snettisham | −1200 | Roger of Rusteni | Wymondham Priory |
| 33. Waxham | c. 1190 | Thomas of Thrune | Hickling Priory |
| 34. Wymondham | −1193 | Earl William II | Wymondham Priory |
| 35. Wymondham | −1193 | Earl William II | Wymondham Priory |
| **NORTHAMPTONSHIRE** | | | |
| 36. Brampton Ash | −1190 | Robert Fitz Adam | Pipewell Abbey |
| 37. Daventry | −1158 | Matilda of Senliz | Daventry Priory |
| 38. Drayton | −1164 × 1197 | William of Drayton | Daventry Priory |

*Continued on next page*

| LOCATION | DATE | PATRON | BENEFICIARY |
|---|---|---|---|
| **NORTHUMBERLAND** | | | |
| 39. Newcastle | −1196 | Robert of Peasenhall | |
| 40. Newcastle | −1196 | Robert of Peasenhall | |
| **OXFORDSHIRE** | | | |
| 41. Cleydon | −1196 | Henry d'Oilli II | |
| **SUFFOLK** | | | |
| 42. Dunwich | −1199 | John of Cove | Knights Templar |
| 43. Elveden | −1200 | Solomon of Whepstead | Bury St. Edmunds Abbey |
| 44. Haberdun | c. 1182 | Dean Herbert | |
| 45. Hienhel | −1198 | Ranulf Dun | |
| 46. Risby | −1200 | R. Fitz Ralph | Bury St. Edmunds Abbey |
| 47. Timworth | c. 1200 | Reginald and Amicia of Groton | Bury St. Edmunds Abbey |
| 48. Tostock | c. 1200 | William Fitz Roger | Sibton Abbey |
| 49. Willingham | −1200 | Robert de Munti | |
| **SURREY** | | | |
| 50. Warlingham | −1200 | Odo of Danmartin | Tandridge Hospital |
| **SUSSEX** | | | |
| 51. Amberley | 1180 1185 | Bishop Seffrid II | Chichester Cathedral |
| 52. Boxgrove | c. 1180 | Roger Hay | Boxgrove Priory |
| 53. Ecclesdon | −1183 | William de Vesci Bishop Seffrid II | Chicester Cathedral |
| **WARWICKSHIRE** | | | |
| 54. Wibtoft | −1189 | Ralph of Arraby | Leicester Abbey |
| **YORKSHIRE** | | | |
| 55. Beeford | c. 1170 | Emelina and Roger of Greynesbury | Meaux Abbey |
| 56. Weedley | −1185 | Nicholas of Hibaldstow | Knights Templar |

12. Post-mill distribution in twelfth-century England. The numbers correspond to those of the entries in Table 1 and the Appendix. Seven installations could not be precisely located: nos. 2, 3, 4, 23, 26, 45, 49.

Because oral testimony was preferred to written documents, many windmills were probably never recorded anywhere. Thus, the countless post-mills that never changed owners, or never were the subject of dispute, are altogether invisible. Despite such frustrating limitations, Table 1 is far longer and more comprehensive than previous summaries of the new technology and would seem to constitute a representative cross section of early English post-mills.

Table 1 suggests that the new technology was introduced before 1137 and was rapidly disseminated after 1180. The Hempnall windmill signals a caution against hasty generalizations, however, because a full sixty-two years separate the possible dates of its construction. Numerous additional post-mills, especially those on small lay fiefs, probably spun their sails before 1200. Indeed, I have encountered references to more than forty windmills that have early-thirteenth-century dates, such as 1203, 1208, 1212, but that may well have been much older.

The oldest windmills were an east country phenomenon. It is as if an imaginary line were drawn from Newcastle to Southampton and all the post-mills were placed east of that divide. Construction of windmills in the other half of the kingdom seems to have gained momentum only after 1200. Environmental conditions were, of course, a major factor. The flat lands, slow rivers, and grain-rich farms of East Anglia welcomed the windmills most enthusiastically, and this region was the mills' favorite habitat. Norfolk attracted the largest concentration. Several mill enthusiasts held properties in different parts of the country, and this dispersion helped spread the use of the new machine.

The mill advocates almost seem to have been united in a tight fraternity of common enterprise. Witness lists reveal that most of them knew several other enthusiasts. This was truly unusual, even in a land as small as England, and it suggests that their friendships and their shared interest in technology were not accidental.

Yet it was not a true fraternity, because women played an important part in this endeavor, as they did in other aspects of the nation's economic life. Four mills were controlled by ladies on their own behalf. Three others were jointly managed by wives and husbands. Some post-mills were acquired by men as part of their wives' dowries, and by women as part of their marriage portions, or through inheritance. Several mills, like the early one at Beeford, had a succession of owners, some of whom were women. Four windmills were given to women's religious houses. One nunnery donated a windmill to a hospital.

Not surprisingly, nine of the known individual promoters and sponsoring institutions owned multiple windmills. Some people became partners in joint enterprises. Several windmill champions had children and grandchildren who built post-mills. All this suggests the existence of an almost missionary zeal with regard to the new apparatus. Windmill supporters also shared a concern for certain religious foundations and health care centers. Most of these individuals were benefactors of hospitals and friends of physicians. Five windmills helped to subsidize hospital activities. Apparently, interest in labor-saving technology and promotion of better medical service went hand in hand.

The first post-mill enthusiasts were overwhelmingly lay individuals. The only clerics were Archbishop Hubert Walter of Canterbury, Bishop Seffrid of Chichester, Abbot Samson of Bury Saint Edmunds, Abbot Roger of Saint Augustine's, Prioress Juliana of Holy Sepulchre, Dean Herbert of Haberdun, and William the Almoner. However, 80 percent of the windmills listed in Table 1 were eventually donated to religious houses. The Benedictine monks acquired nineteen; the Augustinian canons, ten; the Cistercian monks, four; and various communities of nuns, five. The Knights Templar and the secular clergy received two post-mills apiece and the Premonstratensian canons received one. Even if some of the monasteries later rented the wind machines to lay managers, as many surely did, within two generations the bulk of the known facilities ultimately passed to the church.

Ecclesiastical leaders were often fine estate managers, generous landlords, and inventive entrepreneurs, but since religious corporations were inevitably better record keepers than individual barons or small freeholders, more is known about them. Many privately owned windmills never appear in the available documentation and it is therefore entirely possible that lay lords maintained control of many unidentified post-mills.

Most lords, lay and spiritual, were avid builders and commissioned the construction of bridges, canals, castles, and churches throughout the realm. The Norman kings eagerly took the lead in furthering this architectural expansion, but they were surprisingly lackluster patrons of windmill technology. Perhaps these rulers thought that something as common as grain milling was beneath them, or that the machine was basically a poor man's device.

The earliest post-mill was erected on crown land before 1137, but the manor was then held by a churchman working for the government. It was not managed by a royal bailiff. Still, Henry I (1100–1135) and Stephen (1135–1154) could conceivably share some credit for its introduction. Certainly King Henry I and his viceroy, Bishop Roger of Salisbury, were responsible for the peaceful prosperity England then enjoyed. King Henry II (1154–1189) confirmed a subject's windmill donation to Swineshead Abbey and either he or his son Richard (1189–1199) gave the Dunwich post-mill to the Knights Templar. For a price, King John (1199–1216) settled a windmill dispute in Kent. That is the extent of the royal interest.

The lay people who were involved in the utilization of wind power came from several levels of the social hierarchy. The nobility was represented only by an earl, his brother, and a couple of major barons and their mothers. More impressive was the participation of sheriffs and rural knights. They were followed by numerous small landlords, both men and women. There were also estate managers who built windmills on their own holdings, rather than on those of their lord. The typical post-mill proprietor thus belonged to a social stratum approximately equivalent to the middle class today.

Before exploring the experience of the broad range of wind-power promoters, the oldest detectable post-mill merits special attention. Its location seems almost too good to be true—as if a capricious celestial marksman were making sport of this entire study by centering its target as precisely as possible. The locale is further enhanced by an engaging blend of charity and legend.

# Sighting the Target

*I*f one were to shoot an arrow into the skies over England and to pray that it would land in the center of the country, it might well strike earth in Wigston, in southern Leicestershire, the heart of the Midlands. That is the very place where the earliest traces of a windmill were discovered—not on the high hills or the windy coast, not beside a famed cathedral or a towering castle, but deep in the core of the nation. Ironically, there are two villages named Wigston; both hamlets can boast of early post-mills and both can be connected with old King Wiglaf.

Wigston Parva, as the smaller settlement is called, lies about ten miles south of Leicester on the county's southernmost border. It is near the intersection of the ancient imperial highway, Watling Street, with the equally old Fosse Way. The hamlet has so diminished in size over the centuries that it is no longer listed on most maps. Yet in the earliest days of the Roman conquest it was a place of some consequence. Under Emperor Claudius (A.D. 41–54), it was the site of a small, two-acre fort with three gates.

In A.D. 60, this area was the scene of one of the greatest battles in British history. The killing field has not yet been discovered, but somewhere near Wigston Parva at a spot called High Cross, Queen Boudicca made her last stand against the Romans. She and her allies had revolted against horrible mistreatment, thoroughly leveled Colchester and London, and massacred thousands of immigrants. The legions were in Wales at the time, but they raced eastward upon hearing the frightful news. The two armies clashed near Wigston Parva and thousands more died, but the Romans prevailed.

The settlement eventually became a manor of the Saxon and Norman kings and quarreled fiercely with its equally small neighbors, especially about its right to maintain its own parish church. Early in the twelfth century the inventor of the windmill crisscrossed Wigston Parva's haunted fields.

The larger village, Wigston Magna, was a satellite community of a few hundred inhabitants about four miles southeast of Leicester. It was a center of trade and pilgrimage. Although Wigston Magna was nominally controlled by the wealthy earls of Leicester, the villagers retained considerable independence.

"Wigston" apparently means something like Viking's tun, and the name is a reminder of the area's Scandinavian heritage. The spellings of the two Wigstons' names were slightly different in Norman times and probably gave rise to less confusion. What these villages were called before the ninth century has not been determined.

Both vills were recorded in Domesday Book, but neither had a mill in 1066 or 1086. Each had field sites completely unsuited for the exploitation of water power; consequently, early in the twelfth century both erected wind-driven engines. Neither village left records suggesting that its machine was a novel invention; both described their installations merely as mills. Only later, in real-estate charters, were they positively identified as post-mills.

Wigston Parva (*Wicestan*) is only about four miles from Sutton-next-Bosworth (or Sutton Cheney), the manor Sheriff Norman gave to Crowland Abbey and the supposed site of the sheriff's windmill, confirmed by King Wiglaf. In 1485 the area echoed with shouts of praise for the Tudor victory over Richard III. The whole territory had once had a special devotion to Crowland's patron, Saint Guthlac, and in his honor the jurisdictional hundred was called Guthlaxton. Wigston Magna (*Wechingston*) is in Gartree Hundred, but it has even stronger ties to Wiglaf, the Mercian monarch.

Wiglaf's grandson, Wigstan, was killed in 849 in one of the bloody, dynastic struggles that so bedeviled Saxon history. His name was sometimes written Wistan or, more recently, Winston. Wigstan's father may have ruled briefly, or he may have predeceased his son, but Wigstan was too young at his father's death to rule the Mercians. He dutifully asked his mother and a major thane to act as regents. However, he became incensed when the thane, or his son, asked the queen to marry him; Wigstan denounced the proposed union as incestuous. The suitor thereupon had him murdered. The martyrdom, as it was called, occurred at Wistanstow, which seems to be present-day Wistow, not far from Wigston Magna.

When the prince's body was taken to Repton for burial beside his ancestors, it may well have remained overnight at Wigston Magna. Wigstan was locally venerated as a saint, and Repton, Wistow, and Wigston Magna became the centers of his cult. Churches in all three villages were dedicated to him and were the only such shrines in England.[1]

Devotion to Saint Wigstan was strengthened after the Norman Conquest. His bones were moved to Evesham Abbey and subjected to an ordeal by fire to prove their authenticity. Florence of Worcester and William of Malmesbury, two chroniclers who were the first of several to mention him, claimed that Wigstan

---

1. Modern studies of Saint Wigstan include Herbert Thurston and Donald Attwater, *Butler's Lives of the Saints*, 2:440–41; David H. Farmer, *The Oxford Dictionary of Saints*, pp. 410–11; D. J. Bott, "The Murder of St. Wistan"; and D. W. Rollason, "The Cults of Murdered Royal Saints in Anglo-Saxon England," esp. pp. 5–9.

had been beheaded. Pilgrimages were conducted to Wigston Magna, for a bizarre miracle was reported to have occurred annually at Wistanstow: on June 1 of each year, Wigstan's feast day, the saint supposedly let a tuft of his hair grow in the Wistanstow grass for one hour. In 1184 two reputable clergymen were sent to investigate the phenomenon. They informed Archbishop Baldwin of Canterbury that they had actually seen and kissed the holy hair. It is unfortunate that these witnesses did not spend some of their time making technological observations, in addition to conducting pious devotions. There were many windmills to be seen in the area; a couple had even been there for half a century.

Wigston Parva was part of the royal demesne in 1086. William the Conqueror was then using it to benefice a clerk named Aluric. No mill of any kind was cataloged as existing on the property in the exhaustive Domesday survey.[2] Sometime thereafter, the estate passed to William the Almoner. This forgotten clergyman is just now beginning to emerge as an exceptionally important figure in post-mill history. He was the first person to be directly associated with a wind-driven mill.

William was a middle-rank official at the courts of Henry I and Stephen. He served faithfully throughout the thirty-five years of Henry's reign. When Stephen came to power, William continued in his post for about two years, to help ease the transition for the new monarch's staff. A devout priest, he perfected a legible script and was known for his attention to administrative detail. Realizing that these qualities were needed and rewarded in government, he entered the royal service sometime before 1101 as a chaplain and chancery scribe.[3] His ambition and talent must have been substantial, for he quickly assumed greater responsibilities.

By 1106 William had been designated royal almoner. The

2. *Domesday Book*, translated in *V.C.H.*, *Leicestershire*, 1:311.
3. *Regesta*, vol. 2, charters 529 (c. 1101), 598 (1102), 749 (1106). These early attestations are by William the chaplain, almost certainly the same man whose name later appears as William the Almoner, the royal chaplain.

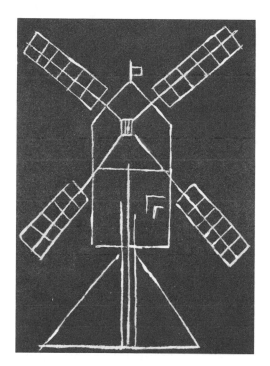

13. Graffito of post-mill from Saint Mary's Church
in Dalham, Suffolk, from a rubbing
made by Violet Pritchard for her study,
*English Medieval Graffiti*. Reproduced with
her kind permission and with that
of Cambridge University Press.
Note the flag atop the mill to indicate the direction
of the wind. The original, one of two post-mills
carved on the opposing responds of the tower arch,
is about ten inches long. It is difficult to date
such depictions, but usually the more
rectangular the sails, the later the windmill.

earliest charter mention of him in that capacity includes the notation that he had prepared the text of the writ himself—an indication that he was still working in the scriptorium.[4] Thereafter his name appears in the available records at widely spaced intervals; he is mentioned as traveling with the king and cooperating with leading members of the court and with his former scribal colleagues.

In 1113 William was at Avranches with Henry I and the latter's physician, Grimbald.[5] About 1121 William and Grimbald attested a royal grant to Harbledown Hospital,[6] as well as a royal writ that concerned milling rights.[7] On the latter occasion William identified himself as both the king's almoner and the king's chaplain—an indication that he served in many capacities.

Sometime before 1130 William attested three different grants for Bernard, one of the chancery scribes. His associates then included Bishop Roger of Salisbury, the former royal chancellor who was then chief justiciar, or viceroy, of the entire kingdom. For decades this genius skillfully managed routine government business, imaginatively reformed financial and judicial procedures, and even administered the whole realm when the king was away in Normandy. Geoffrey, the chancellor; Robert, the keeper of the royal seal; and several other scribes were fellow witnesses.[8] This was a closely knit group of like-minded administrators, all sharing a pride in work well done.

They also shared a desire for wealth and security. By 1130 William had amassed a considerable private estate. Besides the small Leicestershire manor at Wigston Parva, he owned a modest amount of land in Buckinghamshire and more extensive proper-

4. Ibid., charter 643 (1103–1106).
5. Ibid., charter 1015.
6. Ibid., charter 1260.
7. Ibid., charter 1299.
8. John Horace Round, "Bernard, the King's Scribe," pp. 423, 429, 430.

ties in Berkshire and Middlesex.[9] He also had a city tenement in Winchester.[10] In view of his many responsibilities and the necessity that he travel with the restless king, William's visits to Wigston were probably less frequent than were his stops in Winchester and London.

His precise duties as royal almoner are still unclear, but aiding charitable causes was always a significant function of Anglo-Norman government. William must have supervised some such activity, for he was the only bureaucratic official whose title specifically designated him responsible for the public welfare. Additionally, he may have served as an intermediary between the king and some monastic communities.[11] However, it is unlikely that he was consulted about every endowment of a religious institution, or that most monasteries were even considered charities then.

It would be interesting to know the extent of funds allocated to William for direct alms giving. Surely on major ceremonial occasions, like Maundy Thursday in Holy Week, he distributed food, and maybe coins, to the poor. The pipe roll of 1130 reports Exchequer contributions to the infirm, to the indigent, and to special churches in ten different counties.[12] Sometimes funds were designated for even larger projects, such as the construction of bridges, and even of a public lavatory in London.[13] Always, there were scholars wanting stipends and elementary schools seeking

9. The extent of his properties can be estimated from the pardons he received for Danegeld assessments: two shillings in Leicestershire, just about the correct amount for the Wigston holding; twelve shillings in Buckinghamshire; twenty-three shillings in Berkshire; and twenty shillings in Middlesex. *Pipe Rolls, 31 Henry I*, pp. 102, 89, 126, 152.

10. *Winton Domesday*, p. 108, item 538.

11. This function was included in the description of Roger the Almoner's royal duties in 1188; see J. E. Lally, "Secular Patronage at the Court of King Henry II," p. 173.

12. *Pipe Rolls, 31 Henry I*, pp. 12, 24, 44, 76, 109, 126, 131, 135, 137, 141.

13. Edward J. Kealey, "Anglo-Norman Royal Servants and the Public Welfare."

endowments. The number of schools is impressive. During the reign of Henry I at least thirty-three can be identified.[14]

One of the most positive results of public subsidies was an unprecedented rise in health care. The number of physicians increased to such an extent that one doctor was probably available for every one thousand subjects in the kingdom, a ratio many nations would envy today. The surge in hospital construction was even more astonishing. On the average, two new institutions were opened every year throughout the Anglo-Norman period. One estimate suggests that by 1154, one hospital bed was available for every six hundred to one thousand English subjects—this in a period when many illnesses were regularly treated at home.[15]

Financial support for this remarkably widespread endeavor came from all sorts of people. Henry I, and especially his wives, Matilda and Adeliza, were prominent donors. They established some hospitals directly and offered annual subsidies to others. Sixty shillings a year was a favored contribution. A penny a day for each inmate (a higher rate) was also common. Churchmen erected numerous medical facilities—some for the clergy, and many more for the poor, particularly the lepers. Nobles and officials endowed other hospitals. Lesser bureaucrats also joined the movement. A courtier founded a large, multipurpose institution outside London; a chamberlain funded a much smaller one, probably a nursing home, in Winchester; and a royal clerk supported a leper hospital in Shropshire. Women were active patrons; sometimes quite obscure townsfolk joined forces to establish their own hospitals.

When I first studied the development of Anglo-Norman medicine, I was struck by how spontaneously all these individuals

14. Edward J. Kealey, "Schools, Teachers, and Historians of the Reign of Henry I, 1100–1135."

15. These estimates are explained in my *Medieval Medicus: A Social History of Anglo-Norman Medicine*, esp. pp. 49, 88.

aided health care institutions. Now I wonder if there might not have been some inspired coordination of the numerous good works. Not a shred of evidence ties William the Almoner to such supervision. Yet to think that he was not involved in some way is to make him a complete cipher. Such a judgment is totally inconsistent with his three decades of respected service as the king's principal agent for charitable concerns. It seems more logical to view William as something of a voiceless visionary, an individual who quietly did great things but let others take the credit. A reflective person, he must have wondered if there were not some way to lighten the load of toiling people and to provide more food to the increasing population. Eventually he found such a device: the windmill.

The machine must have been known in King Henry's time and it should certainly be considered another exciting achievement of his era. However, that monarch's death in 1135 and a consequent redirection of William the Almoner's vocation were required for the post-mill to appear in contemporary records. The true significance of the copyist's terse entry would not be realized for many centuries, however.

The king's death brought many changes, especially in his household and court bureaucracy. Some officials, such as Robert de Sigillo, the master of the royal seal, soon retired. Others, William included, persevered for a few years. He prudently lingered just long enough (until 1137) to help others and to benefit from the new king's first fickle show of largess. Thereafter, governmental service became distinctly unpleasant for him.

In 1139 Stephen viciously turned against his mighty viceroy, Bishop Roger of Salisbury, and forced him to surrender his power.[16] Consequently, more than half the scribal clerks hurriedly left the government. Their action seriously disrupted many ad-

---

16. Edward J. Kealey, *Roger of Salisbury, Viceroy of England*, pp. 173–208.

ministrative functions for the remainder of Stephen's reign.[17] It is not clear whether anyone even succeeded to William's post.

William was over sixty years old in 1137 when he decided to abandon the secular world entirely and to answer a higher call, to monastic life at Reading Abbey. This vigorous new house was a particularly fitting choice. It had been founded by King Henry, and he had selected its church for his final resting place. It was also the cloister that William's chancery friend, Robert de Sigillo, had entered. The first abbot, Hugh, had been a learned leader of international repute. The second, Abbot Anscher, was so distressed at "the large numbers of various infirmities among the poor at that time" that he established a leper hospital in Reading and drew up a set of statutes specifically regulating its management. The abbey also became very active in support of local education.[18]

Another chancery appointee, whom paleographical analysis reveals only as Scribe 13, was also associated with Reading Abbey. He had a pretentious writing style and a somewhat inconsistent pen, but he had transcribed numerous writs for the two kings. He left the administration about the time of Bishop Roger's disgrace and soon began to prepare documents for the abbey. When Robert de Sigillo was promoted to the bishopric of London in 1141, Scribe 13 wrote out a charter for him. Oddly enough, he seems for a while to have become part of a ring of forgers who drafted monastic charters and then appended to them a fraudulent first seal of Stephen, thereby trying to prove the grants were both genuine and made before 1139.[19]

King Henry was a demanding, sometimes cruel, master, but

---

17. Edward J. Kealey, "King Stephen: Government and Anarchy."

18. B.L. Egerton MS. 3031 (Reading Abbey Cartulary), fol. 11v; Cotton MS. Vespasian E V (Reading Abbey Cartulary), fols. 38–39; printed in Dugdale, *Monasticon Anglicanum*, 4:42–45. In *Medieval Medicus*, p. 202, I expressed an interest in editing these statutes, but Dugdale's version makes that unnecessary.

19. T. A. M. Bishop, *Scriptores Regis*, plate 17a; *Regesta*, 3:xiii–xiv; ibid., vol. 4, plates 5–11 on pp. 5–11; Kealey, "King Stephen," pp. 208–9.

he inspired genuine affection among many of his subordinates, and some of them gathered round him in death as faithfully as they had served him in life. Their devotion is a tribute to the tone of government set by Bishop Roger (whom contemporaries referred to as "second only to the king"[20]), as well as to his administration's attempt to respond to social concerns more effectively than had its predecessors.

When William the Almoner became a monk, he decided to make a financial contribution to his new home, as was customary for wealthy recruits. He chose to give Reading Abbey one of his holdings, that in Wigston Parva. Since his other lands were never individually named, it is now impossible to trace their disposition. Presumably, some of them reverted to royal control. The Winchester tenement, however, had been reduced to waste by 1148 and was then paying rent to the diocesan bishop. It was still remembered as having been owned by William the Almoner—a minor echo of his reputation among the townsfolk.[21]

Wigston Parva, of course, was crown property. To deed it to Reading Abbey William sought, and obtained, the consent of King Stephen. Someone was rather careless with the king's charter to William, however. A scribe wrote a heading and left space for the rest of the charter to be entered in the abbey cartulary, but the text was never copied.[22] That omission is a major loss to this investigation.

Early in his reign the king desired to honor his late uncle, Henry I. He therefore made a number of gifts to Reading Abbey and, in 1137, officially granted to the monks William's small estate at Wigston Parva, *including its mill.*[23] Stephen's brief an-

20. Kealey, *Roger of Salisbury*, pp. 26–81.

21. *Winton Domesday*, p. 108, item 538.

22. B.L. Egerton MS. 3031 fol. 18v. The space was reserved in Harley MS. 1708 (Reading Abbey Cartulary), fol. 30v.

23. B.L. Egerton MS. 3031, fol. 18v; Harley MS. 1708, fol. 30; Cotton MS. Vespasian E XXV, fol. 24; printed in *Regesta*, 3, charter 681. The Egerton scribe indicated

nouncement, so long unnoticed in the records, is the earliest detectable reference to a post-mill. The king's language was ambiguous, but other records clearly indicate that the manor had only one mill. By the end of the century, it would be described more explicitly as a wind-powered installation, but the charter recording Stephen's gift is proof that the machine was operating before 1137.

It is extremely satisfying to nominate William the Almoner as the premier windmill advocate. He had successfully balanced his responsibilities to church, state, and commonweal and was an appropriate pioneering champion of labor-saving technology. The possibility that the two kings also played a role cannot be dismissed entirely, however. As the ultimate lords of the Wigston manor, they may have offered some encouragement for development of the new process. William's involvement ceased when he became a monk. He wanted to retire to the peace of monastic seclusion and was content to let the windmill stand on its own merits.

The mundane affairs of the estate demanded someone's attention. Therefore, shortly after 1137 the Reading monks decided to rent Wigston manor, rather than to manage it directly. Evidently, William had failed to convince his brothers of the full value of their new mill, and they failed to realize that it would be more profitable to keep the apparatus under their personal control. The result of the abbey's decision was that administration of the manor was thenceforth largely influenced by feudal politics.

For the remainder of the century a succession of lay landlords administered the manor at Wigston Parva. They soon became involved in regional affairs, particularly those of the neighboring hamlets, Claybrook and Wibtoft. Thereafter the three settlements shared prowess in development of the novel technology and were

---

that Stephen had issued two other charters concerning William's Wigston gift, but they were never copied.

the subjects of an awkward ecclesiastical arrangement whereby Claybrook became the dominant parish and the other two were reduced to dependent chaplaincies. The residents of Wigston had always seemed to fear being overwhelmed by a larger village and they now took steps to preserve their settlement's separate identity. As a last resort, they could always call upon their ultimate landlord, Reading Abbey.

The most important feudal magnates in the area were the Beaumont earls of Leicester (who in the twelfth century were unimaginatively named Robert I, II, III, and IV) and their close relatives, the earls of Warwick. The members of both families were warriors, courtiers, crusaders, patrons of monasteries, friends of physicians, allies of windmill pioneers, prolific authors of charters, and participants in numerous milling initiatives. The stewards, or estate managers, of the Leicester earls, repetitively named Arnold Wood (Ernald de Bosco) I–IV, were, like their lords, bustling, ambitious men. Their entourages, and those of the earls, included a large number of doctors and other university graduates. Such cosmopolitan individuals contrasted sharply with the quiet rural farmers who constituted the majority of the area's population.

Wigston Parva was first rented for thirty shillings a year.[24] That was ten shillings more than the manor had yielded in 1086; the difference may represent the estimated value of the new mill. Relations between the monks and Arnold Wood III, the first tenant, were not altogether pleasant. Something went wrong, for the steward returned Wigston manor to direct monastic control sometime between 1148 and 1155.[25] Perhaps the monks feared

24. B.L. Egerton MS. 3031, fol. 44; Harley MS. 1708, fol. 121; Cotton MS. Vespasian E XXV, fol. 66v. This is a charter between Earl Robert II of Leicester and Abbot Edward (1136–1154) guaranteeing Arnold's agreements with the abbey.

25. Ibid.; this information was copied in all the manuscripts. Arnold noted that he was to hold the land only as long as it pleased the abbey. His reversion was witnessed by Earl Robert of Leicester and by Robert de Belemis, the prior of Bermondsey (1148–1155).

that their small estate would be swallowed up in Arnold's surrounding holdings, for he had already acquired Claybrook and parts of Wibtoft.

Concurrently, King Stephen had to order Arnold's men not to make any claim on Wigston Parva.[26] Moreover, the abbey obtained confirmations of its possessions from all the major authorities—the earl of Leicester, the bishop of Lincoln, the archbishop of Canterbury, the new monarch, Henry II, and a succession of popes.[27] These general proclamations listed many of Reading Abbey's holdings but referred to this estate as the land of William the Almoner. Unfortunately, they did not include any specifics about its mill.

After the troubles with Arnold had subsided, the abbey leased Wigston manor to Master Gilbert of Belgrave. This prominent physician from the suburbs of Leicester may have been chosen as a result of negotiation with Arnold. The steward and the doctor subsequently cooperated in a number of commercial ventures, particularly in Claybrook whose control they eventually shared. It is likely that this practitioner was the Master Gilbert who gave a volume of decretals, and perhaps two sets of scriptural glosses, to

26. B.L. Harley MS. 1708, fol. 30; calendared in *Regesta*, 3, charter 682 (1148–1154). Pope Alexander III (1159–1181) wrote to the bishops of London and Chichester about the squabbles between the churches of Wigston and Claybrook; Egerton MS. 3031, fol. 75.

27. B.L. Egerton MS. 3031, fols. 21–21v; Harley MS. 1708, fol. 21. These manuscripts include a confirmation by Henry II regarding the gifts of his grandfather, including Wigston, "which was the land of William the Almoner." No mention is made of King Stephen, indicating that this was a rather inaccurate, mean-spirited confirmation. This is the only suggestion that Henry I had anything to do with the Wigston grant to Reading. If it is true, the post-mill predates 1135. Henry II's charter, written in Rouen between 1156 and 1157, still exists; see B.L. Additional Charter 19571; reproduced in G. F. Warner and H. J. Ellis, *Facsimiles*, charter 40. Confirmations by Kings Richard and John are found in Harley MS. 1708, fols. 31, 34; those by Popes Eugenius III, Adrian IV, and Alexander III are found in Egerton MS. 3031, fols. 65v, 66v, 68. On Bishop Robert, see D. M. Smith, ed., *Episcopal Acta*, 1:144, charter 230; on the archbishop, see Avrom Saltman, *Theobald, Archbishop of Canterbury*, charter 215.

the Reading Abbey library. The gift may well have been an attempt to smooth over past disagreements.[28]

Doctor Gilbert is reminiscent of a character in *Ivanhoe*. He was an ambitious, reckless person who had powerful friends and ample, if borrowed, cash. In 1179–1180 he became involved in a quarrel that could only be settled by a judicial duel, or trial by combat. He hired a champion to fight for him, as the law allowed, but that bruiser unaccountably withdrew. Gilbert then had to pay the Exchequer a five-mark fine, a bit more than the normal forfeit. It took him only three years to clear the debt. Soon thereafter he borrowed more than eighty pounds sterling (quite a hefty sum) from a Jewish moneylender. This man, the well-known Aaron of Lincoln, also held land in Belgrave—an unusual privilege for a Jew at the time. Like many of Aaron's debtors, most of whom were also small landowners, Gilbert borrowed to finance his investments. He wanted venture capital and evidently used it to enlarge his holdings in Claybrook, presumably with the acquiescence of Arnold Wood and his family.

When Aaron died in 1186, his property and the debts owed him passed to the government. It was normal for the crown to demand a third of a deceased Jewish person's estate, but Aaron's resources were so enormous that the king decided to confiscate them all. Ironically, the bulk of the treasure seized at Aaron's death was lost at sea in 1187. Unpaid debts were relentlessly collected, however. Most of the outstanding obligations were paid

28. His tenure at Wigston is inferred from the facts that he held other area properties (some jointly with his son, Peter); that his son later leased Wigston (but he was only a boy in 1155); and that Petronella of Belgrave asserted a claim to Wigston by right of *mort d'ancestor* in 1227, when Peter was still alive (see footnote 33 to this chapter). On Gilbert's transactions with regard to Claybrook, see B.L. Additional Charters 47398, 47551, 47552, 47556, 47557, 47558, 47573R, 47573T, 47573U, 47580; see also Innocent III, *The Letters of Pope Innocent III (1198–1216) Concerning England and Wales*, p. 134, letter no. 814. On the book donation, see B.L. Egerton MS. 3031, fols. 8v–11; Samuel Barfield, "Lord Fingall's Cartulary of Reading Abbey," p. 117.

into a special "Aaron's Exchequer" until 1191. Thereafter, the royal Exchequer pursued the remaining moneys directly, sometimes for decades on end.

Doctor Gilbert received especially favorable terms for his loan; he was required to pay only a pound a year against an eighty-pound balance. At that rate he would carry his debt for an infinity. He did, in fact, continue to liquidate it in tiny amounts until his death in 1214, whereupon his son, Peter of Belgrave, assumed the payments. Gilbert could not resist the temptation to try to evade part of his outstanding obligation. More than once, he improperly reduced, or understated, his balance. Each time the Exchequer tripped him up and reminded him to pay in full.

Although a real-estate speculator, Gilbert was also genuinely interested in certain aspects of his medical practice. He conferred with a number of other physicians, including the learned author Master Gilbert of Aquila. Gilbert of Belgrave's personal seal still exists, attached to a Claybrook agreement he and Arnold Wood IV concluded with the prioress of Nuneaton. It is the oldest extant signet of an English physician. The design is partly conventional and partly unique. The legend (which reads "The seal of Master Gilbert") encircles a naked man in an odd, distorted pose whose arms and legs are extended like those of an Oriental dancer. Could he be simulating a set of windmill vanes?[29]

Gilbert seems to have obtained Wigston manor largely as an early endowment for his son, Peter.[30] Together they also admin-

29. On Gilbert's duel, see *Pipe Rolls, 26 Henry II*, pp. 100, 102; *27 Henry II*, p. 75; *28 Henry II*, p. 94. On his debt to Aaron the Jew, see *Pipe Rolls, 3 Richard I* through *16 John*. On Gilbert's attestation with Master G(ilbert) of Aquila, see B.L. Additional Charter 47551. His seal is attached to Additional Charter 47559.

30. B.L. Additional Charter 47560. Peter, who became a knight rather than a physician, was summoned to fight with the feudal host in Poitou in 1214; see *Pipe Rolls, 17 John*, p. 103. He continued to pay his father's Exchequer debt after that until at least 1242; *Pipe Rolls, 26 Henry III*, p. 179. Peter was also bequeathed a parcel of land in Watton, as well as its unspecified mill, by a William of Belgrave. Peter gave it to a Ralph the janitor, who gave it to Stephen of Seagrave; see John Nichols, *The History*

istered parts of Claybrook. It is not known how long Peter held Wigston, but after the death of his nephew Eustace, he decided to return the manor to the Reading monks.[31] Because he had granted the small part of the hamlet where the mill was located to Ralph of Arraby, he had to convince Ralph to surrender his holding to the abbey as well.

Sometime before 1200, Ralph agreed that the monks should possess the whole windmill (*totum molendinum ad ventum*), its site in the Wigston field, and the strips of land that pertained to it. All of these he reported having received from Peter of Belgrave. His declaration is the first known instance in which the mill of Wigston Parva, which was then more than sixty years old, was explicitly identified as a post-mill.[32]

Ralph's grant to Reading Abbey was made without reservation, but it was not characterized by willing generosity either. In fact, the abbot and monastery had to pay him twenty silver marks for cession of the windmill. This was a substantial sum and must represent a type of sales price, or valuation of the mill. Apparently, the monks had finally realized that it was in their interest to supervise the mill operation themselves. The issue was not completely settled, however, for there was some subsequent squabbling among the Belgrave heirs. As late as 1227 the abbey paid Petronella of Belgrave, who was presumably Gilbert's

---

*and Antiquities of the County of Leicester*, vol. 2, pt. 1, p. 113. William of Belgrave attested for Abbot Paul of Leicester between 1188 and 1204; B.L. Additional Charter 47580; printed in Frank M. Stenton, ed., *Documents Illustrative of the Social and Economic History of the Danelaw*, charter 321. He was also present at litigation in 1200; Doris M. Stenton, ed., *Pleas before the King and His Justices, 1198–1212*, vol. 1, pleas 2869, 3004.

31. B.L. Harley MS. 1708, fol. 121v; Cotton MS. Vespasian E XXV, fol. 67. The return of Wigston to Reading is not mentioned in Egerton MS. 3031, which was composed in Richard I's reign; therefore, it may have happened in the late 1190s, or even early in the reign of King John. Peter was still paying thirty shillings a year for the manor, a reflection of price stability and of the power of tradition.

32. B.L. Harley MS. 1708, fol. 121. In the Cotton MS. Vespasian E XXV, fol. 67, the red letter heading indicates it was a declaration of Rich(ard) of Arraby, but in the text the name is correctly copied as Ralph. Evidently the scribe was nodding a bit.

daughter, sixty marks to quitclaim her rights to the Wigston manor.[33]

Because Ralph of Arraby declared that he was giving the abbey the charter that Peter of Belgrave had given him, as well as all his other documents pertaining to the mill, there ought not to have been any subsequent problems, at least about the windmill. The Reading scribe certainly intended to copy Peter's writ to Ralph, for he reserved space and wrote a heading for it in the cartulary, but the text was never entered.[34] Since this had happened before, something was clearly wrong with the abbey's archival procedures. Most likely, the former royal clerks who had become monks had died by the time these mistakes were made.

This pioneering post-mill created an unusual amount of parchment traffic. The lost charters would have been invaluable for this study, but a large number of others were preserved. The monks obviously wanted complete records in order to forestall further litigation. Their hopes were frustrated, and this fact probably worked in Petronella's favor.

Ralph of Arraby owned other mills in addition to the Wigston post-mill that he had rented for a while. To provide for his eventual burial in their church, he gave the monks of Leicester Abbey two mills in Wibtoft, a horse mill (for which animals provided the power), and a windmill. Arnold Wood confirmed the grant and Henry II mentioned it in one of his proclamations.[35] This means the post-mill definitely existed before 1189.

Ralph of Arraby probably knew, or had connections with, a few other post-mills. The third hamlet in the triangle formed with Wigston Parva and Wibtoft was Claybrook and Ralph attested one of Arnold Wood's writs pertaining to it.[36] That village

33. B.L. Harley MS. 1708, fol. 177v. Petronella obtained a writ of *mort d'ancestor* to press her claim.
34. Ibid., fol. 122v.
35. See the Appendix, entry no. 54.
36. B.L. Additional Charter 47573T (c. 1160–1170).

had a water mill and a windmill by 1279, at the very least.[37] There was probably an early post-mill at Arraby as well. In the late twelfth century, some people there became anxious that an unspecified mill might collapse or malfunction.[38] This would seem to be worry about a new type of machine. Other members of Ralph's family, probably in the next generation, also became windmill sponsors.[39]

The original Wigston Parva facility was clearly a post-mill. Throughout the hamlet's Norman and Plantagenet history, only one mill was ever mentioned as existing there. No mill of any kind was listed for Wigston in the exhaustive Domesday survey, either for 1066 or for 1086. Beginning in 1137 the village's records consistently include mention of a single installation; therefore, it was erected between 1086 and 1137.

In the latter part of the twelfth century, that one mill was explicitly referred to as a windmill; later in the same charter, it was called a mill. There was still only one mill in Wigston Parva in 1279, but then it was precisely identified as a windmill.[40] The conclusion is as inevitable as it is welcome: the late-twelfth-cen-

37. Nichols, *History and Antiquities*, 4:104.

38. Ibid., vol. 1, pt. 2, appendix, p. 90. Croxton Abbey had received the mill from Robert of Arraby (perhaps Ralph's brother), who also gave a church to Leicester Abbey during Henry II's reign: ibid., p. 58; Dugdale, *Monasticon Anglicanum*, 4:468. He also gave the Leicester monks an unspecified mill at Watton; see Nichols, *History and Antiquities*, vol. 1, pt. 2, appendix, p. 113.

39. The nephew of Robert of Arraby II, William Boteler, gave the Premonstratensian canons of Croxton Abbey (founded before 1159) free access to a windmill in Eaton, in northern Leicestershire. It seems to have belonged previously at least to Robert II and was built next to a fulling mill; see Nichols, *History and Antiquities*, vol. 2, pt. 1, appendix, p. 91. There was considerable protracted dispute about the tithes on the profits of the Eaton mills and about other abbey property there. In 1230 the abbot of Croxton made an agreement with Henry of Seagrave, a relative of the Arraby family, giving him the windmill at Ilston-on-the-Hill, near Wigston Magna; ibid., vol. 2, pt. 1, p. 172, and appendix, pp. 89, 94. That post-mill and the one at Eaton may have had a long life, but they cannot be definitely dated before 1200 and therefore are not included in Table 1.

40. Ibid., 4:124, from an inquest of 7 Edward I. A single post-mill was also mentioned in a record of 1296; ibid.

tury post-mill is the same apparatus as the one of 1137; it is the oldest of the known windmills.

The terminology is obviously inconsistent, but as previously mentioned, such inconsistency was not unusual. The Reading Abbey cartularies and the charter copies clearly reveal an interchangeability of the words "mill" and "windmill." The rubric of Ralph of Arraby's charter refers to the mill of Wigston; in the body of the text, the word "windmill" appears once and the vague term "mill" appears three times. The heading of Peter of Belgrave's lost charter simply reads "mill."

The Wigston Parva gift occasioned no excitement among contemporaries. It was mentioned in a calm, matter-of-fact manner; evidently no one considered it a novelty. The lack of fanfare about its appearance suggests that the invention had made its debut at least several years (perhaps as many as twenty) earlier. The point, of course, is that Wigston is the earliest windmill whose existence has been discovered in the records. It is not necessarily the premier windmill.

But what about other places where the arrow might fall, such as Wigston Magna? This village ten miles northeast of Wigston Parva had documentary and environmental similarities to the smaller hamlet, but its tenurial factors were far more complicated. It, too, had no mill in 1066 or 1086 but acquired one long before 1169, and many years would pass before that unspecified mill would be identified properly in medieval records.

Wigston Magna was not a royal village that passed to monastic control but rather one of the many manors belonging to the wealthy earls of Leicester. The settlement was subdivided among dozens of staunchly independent peasant freeholders. Neither the church nor the royal court had any significant property there, except for some minor grants made by the earls to Lenton Priory and Brackley Hospital.

Like a few other Leicestershire villages, populous Wigston Magna was actually a double vill, two separate communities, each with its own church and agricultural strips. Eventually, there were several mills, including a water mill that still exists in the lower section, and a few upland post-mills, whose abandoned mounds now dot the open fields. The early records that survive pertain only to the upper section (the settlement built around Saint Wigstan's Church), an area totally unsuited for the use of water power.

The existence of a windmill in the upper community is mentioned in thirteenth-century texts, but its date can be pushed back several generations, to a period well before 1169. Sometime prior to that date, the earl of Leicester confirmed the holdings of Simon of Wykingeston, including a mill, an acre of land, and paths to the mill, all of which he had received from his father, Lawrence, and from Reginald Winterborn. They had received these holdings from the Englishmen in Wigston, and for them they paid the earl only a penny a year—surely a nominal fee. This cooperative sponsorship of a mill by a group of Englishmen is extremely interesting. Presumably the villages both erected and operated the mill as a communal enterprise.

The earl's charter for Simon was witnessed by a dozen men, including several local dignitaries. The original, which may have had a more precise identification of the mill, has been lost. The text can be found in a fifteenth-century reproduction of the charter to which the earl's genuine seal, now damaged, is attached.[41]

41. Alexander H. Thompson, ed., *A Calendar of Charters and Other Documents Belonging to the Hospital of William Wyggeston at Leicester*, p. 451, charter 867. The editor dated the charter c. 1170–1180, but that estimate can now be made more precise. Earl Robert II ruled 1118–1168; Robert III ruled 1169–1190. The witnesses to the charter between the earl and Simon included Thurstan, abbot of Garendon (c. 1155–1180); Robert, prior of Kenilworth (1158/59–1186); Hugh Barre, archdeacon of Leicester (from about 1148 until 1157/58, when he resigned, but he lived for a while thereafter); William Basset, sheriff of Leicestershire (1162/63 to 1168/69); and William the physician. Several of the witnesses also attested a grant by Robert II to Nuneaton Priory, definitely datable to 1162–1168; B.L. Additional Charter 48086. W. G. Hoskins, who analyzed Wigston Magna extensively in his valuable *The Midland Peasant: The Economic and Social History of a Leicestershire Village*, mentioned the charter and emphasized that the available early documents related only to the upper section but that the Sence, a stream in the southern part, was the only body of water that could have powered a water mill (see pp. 7, 26). Nevertheless, he felt the charter must refer to a water mill. He was probably another eminent authority who was unconsciously wedded to the long-held scholarly belief that windmills did not exist before 1185.

14. Windmill drawn by an East Anglian artist
about 1340. From Additional MS. 42130
(The Luttrell Psalter), fol. 158;
reproduced by permission of the British Library.

The references to paths to the mill and to the one-acre mill
site are characteristic of windmill, not water mill, charters. However, this facility can be positively identified as a post-mill because
of the waterless terrain on which it was situated. In addition, a
survey of upper Wigston in 1296, which indicated that it then
had only one mill, a windmill, substantiates that conclusion.[42]
This would have been Simon's mill, or its direct descendant. One
interesting fact revealed by this investigation is the startling longevity of the mill structures. Moreover, once they aged beyond

42. Hoskins, *The Midland Peasant*, p. 26.

repair, a successor mill was usually erected on exactly the same spot. A descendant of the twelfth-century Friskney post-mill in Lincolnshire, for example, is functioning there in the twentieth century.

Since the Wigston Magna post-mill had two sets of known owners before 1169, obviously it had operated for a considerable time—surely at least ten years—before that. The cooperative sponsorship by the English residents also suggests that the windmill was quite old. Whether the "English" here meant Saxons as opposed to Normans, or perhaps Saxons as opposed to those of Scandinavian descent who were predominant in this former Danelaw area, cannot be determined.

Cooperative community initiatives can be seen in another midcentury charter for Wigston Magna. The names of the witnesses reveal that the settlement (here perhaps the combined vill, rather than just one of its sections) had two clerks, a sergeant, a chaplain, and an official in charge of the village's chamber, or treasury.[43] This charter offers an unusual glimpse into the organization and vitality of village life. Arnold Wood, the earl's steward, also attested the transaction. It is tempting and altogether plausible to cast him as the herald who explained the technological breakthrough.

The two Wigstons thus reveal the origins of the windmill and exemplify the polar extremes of its sponsorship. The mill at Wigston Parva was the result of one successful official's educated vision. That at Wigston Magna was cooperatively managed by a group of modest, thoroughly obscure individuals. Both models would be emulated in the next generation.

There would later be a veritable explosion of windmill construction in Leicestershire. Besides the post-mills at the two Wigstons and at Wibtoft, one existed at the Augustinian Priory

---

43. Thompson, *A Calendar of Charters*, p. 452, charter 868. The last individual was called Richard of the Chamber (*thalam*) of Wikingeston.

of Bradley, certainly before 1220.[44] Croxton Abbey acquired early windmills at Eaton and, before 1230, at Ilston-on-the-Hill, which is just down the road from Wigston Magna.[45] Post-mills were also erected at Claybrook, a village under the control of the de Bosco stewards, and at Bitteswell and Sharnford. All three are almost within hailing distance of Wigston Parva. These post-mills appear only in a 1279 Leicester survey, along with at least twenty other wind-powered installations.[46] Twelfth-century dates cannot be assigned to them with any certainty, but some seem to have been in operation for a considerable period. Other early sites in Leicester, such as the mounds of the two abandoned post-mills at Kirkby Bellars, were never recorded at all (see Figure 9).

There was also a pre-1279 windmill at Sutton-next-Bosworth, the ancient site of Sheriff Norman's manorial gift and the supposed location of the windmill confirmed by Wiglaf. How sad that it cannot be given an earlier date.

It is worth asking at this point why this novel invention appeared when it did. After all, more than a millennium had passed since the invention of the water mill, humanity's first truly complex machine. What were the conditions existing in the years immediately before 1137, and in Norman England, that inspired the second productive mechanical apparatus? Except for the lengthy Investiture Controversy (1059–1122), which concerned the relative primacy of spiritual authority and temporal power,

44. *V.C.H., Leicestershire*, 2:24. The abbey had owned the windmill since its foundation, but that date cannot be fixed, except that it was sometime before 1220.

45. See footnote 39 to this chapter.

46. I did not make a systematic search for thirteenth-century Leicestershire windmills, or for post-mills existing anywhere after 1200. However, Nichols, in *History and Antiquities*, mentions the existence of windmills in the following locations before 1279: Cosby, Huncote, Ibstock, Shilton, Nailstone, Orton, Sapcote, Glenfield, Stony Stanton, Sutton Cheney, Bagworth, Thurleston, Obaston, Normanton Turbille, Miserton, Poultney, Swinford, and Willoughby Waterless. Thompson, in *A Calendar of Charters*, p. 239, charter 355, mentions a mill at Foxton before 1295, and C.U.L. MS. Dd. 3.87 art. 20 (Owston Abbey Cartulary), fol. 24, includes mention of a mill at Owston before 1293. Ranulf, earl of Chester from 1187 to 1232, gave Leicester Abbey both a windmill and a water mill at Barsby, Leicestershire, but the grant cannot be positively dated before 1200; Oxford University, Bodleian Library Laud MS. Misc. 625, fol. 20.

England enjoyed domestic peace, innovative governmental administration, and widespread economic prosperity from the accession of Henry I in 1100 until King Stephen's repudiation of Bishop Roger of Salisbury in 1139. Literacy had increased markedly in England, and on the Continent, among both the clergy and the laity. Some of the new learning was an accidental by-product of the intellectual controversies over church reform. Theology and law had always been the premier studies, but by 1100 scholars in other disciplines were building reputations throughout Christendom and the rudiments of university education were beginning to be defined. Long-neglected fields such as astronomy, mathematics, and medicine began to attract more attention. Even merchants and governmental officials self-consciously examined their own specialties.

Nationalistic fervor was expanding personal loyalties, and many restless souls idealistically joined vast international crusades to redeem the Holy Land. Solid, but impressive, Romanesque architecture was exploring dazzling technical and aesthetic possibilities. Tales of courtly love were challenging the primacy of heroic epics. New lands were being cultivated, and temporary market fairs were yielding to permanent free towns. Kings and nobles were augmenting their power, but the son of a serf could become pope, and women were exercising unprecedented influence in local affairs.

Moving in tandem were the increasing economic independence of some skilled workers and a heightened appreciation for all manual arts. Voluntary physical labor had played an honored part in Benedictine monasticism for centuries, but individuals who were compelled to toil for a living had had little genuine freedom of action. They usually found themselves bound as a result of heredity to an arbitrary lord who dictated the most minute aspects of their existence. By the early twelfth century, countless former serfs had fled their burdensome manorial obligations to become free citizens in the new towns that were springing up everywhere. Most people remained on the farms, however.

In some rural villages those peasant families who had quietly managed to retain their free status over the centuries sought increased opportunities to express themselves more openly.

Many average people were discovering dignity and personal satisfaction in their daily occupations. In writ after writ, ordinary individuals proudly identified themselves as engineers, masons, carpenters, merchants, cooks, scribes, physicians, and millers. Craftsmen and merchant adventurers regulated their trades and formed fraternal associations for social and business purposes, including insurance protection. These powerful new guilds enhanced the attraction of urban life, promoted technical innovation, and maintained high professional standards by requiring the presentation of a masterpiece for full membership certification. Manuscript illustrators began to treat secular subjects, such as historical events and medical procedures. About A.D. 1000 God the Creator began to be depicted as a master craftsman holding scales and an architect's compass.

The church still extolled the values of monastic solitude and many lay people were attracted to the cloister at different stages in their careers. However, some intellectuals preached the need for more direct involvement in worldly affairs. On the Continent an early-twelfth-century monk who had chosen the pseudonym Theophilus wrote the first original medieval treatise on a craftsman's art. He discussed techniques of glassmaking and metallurgy, the mechanical crank, the use of the flywheel to equalize rotation, and the process of drawing wire. Theophilus also explained that man's dignity was exhibited not only by his rationality but also by his ability to perfect craft skills that could be used for the ultimate service of God. At Paris in the 1120s, the erudite canon Hugh of Saint Victor also defended the place of the mechanical arts in the totality of human knowledge.[47]

47. For overviews of the emphasis on mechanical arts, see John Van Engen, "Theophilus Presbyter and Rupert of Deutz: The Manual Arts and Benedictine Theology"; George Ovitt, Jr., "The Status of Mechanical Arts in Medieval Classifications of Learning."

Honorius Augustodunensis, an encyclopedist and biblical popularizer who lived in England and Germany, also emphasized this theme. Pedro Alfonso, the Spanish physician, scientist, and witty raconteur who converted from Judaism to Catholicism in 1106 and thereafter spent several years in England, stressed the importance of practical learning. His associate, Adelard of Bath, was intrigued by mathematics and natural science. Both he and Pedro urged people to observe, to test, and to discover the true causes of things.[48] The garrulous churchman Gerald of Wales, reflecting on the inventive process, claimed that many of nature's riches still lay hidden because man had not given them due attention. Earlier generations had been driven by "sheer urgent necessity" to invent some of the "amenities of human existence"; zeal and industry had brought supplementary advantages to their successors.[49]

Along with such unprecedented praise of nontraditional learning, there was a naive delight in gadgetry. Every twelfth-century naturalist worth his salt wrote a treatise on the astrolabe. Many mathematicians and scientists explored the wonders of the abacus; the bureaucrats at the Exchequer adapted its principles for their own use. Innumerable pages of parchment were devoted to complex tables facilitating astronomical, and astrological, prediction. Churches boasted simple water clocks, but their complex mechanical pipe organs really were wonders.

Anglo-Norman administrators acutely analyzed their own profession. Their passion for statistical surveys and accurate records was unequaled in Continental Europe—witness their compilation of Domesday Book. They also wrote histories, biographies, legal codifications, and perceptive analyses of their own times. In 1135 an anonymous clerk outlined the organization of

48. On Adelard, Honorius, and Pedro, see Kealey, *Medieval Medicus*, esp. pp. 13–15, 75–81.

49. Gerald of Wales, *The Journey through Wales and the Description of Wales*, p. 116. I owe this reference to Alice Hynes, "The Mill in England, 1066–1307: The Influence of Technology on Society."

the royal household, including the duties of each member. This management treatise embodied a recognition of the distinct contribution of minor employees, such as cooks and chamberlains.[50]

The self-confidence and vocational pride expressed by such writers excited many artists, craftsmen, and engineers, whose creations had almost invariably been designed to serve the pleasure of someone else, usually a clergyman or noble. The increased esteem for labor stimulated a new curiosity and inventiveness among working-class people and encouraged them to devise more personally useful creations.

Not content to passively await change, English people from all walks of life courageously decided to alter their circumstances. Despite their traditional respect for hierarchical position and class privilege, many men and women were anxious to improve the general welfare. Their varied projects influenced each other in subtle but substantial ways. Social concern was evidently almost as important to development of the windmill as technical expertise, for the engine was essentially a labor-saving device, a means for transforming free energy into useful work.

Both the burghers and the independent farmers jealously guarded their hard-won political liberties, but they also fought hard to preserve something of the cooperative spirit that characterized rural life. The strongest indication of Anglo-Norman community interest was an unprecedented drive to construct health care institutions—hospitals, leprosariums, and retirement homes—throughout the country for people of all ranks, but especially for the poor. The effort began early in the twelfth century and never let up. A somewhat similar campaign led to the rapid multiplication of elementary schools, orphanages, and travelers' hospices.[51]

Thus, people throughout Europe, but most especially in En-

50. The *Constitutio Domus Regis*, edited and translated by Charles Johnson, is included in his edition of Richard Fitz Nigel's *The Course of the Exchequer*, pp. 128–35.
51. For a discussion of health care in Norman England, see Kealey, *Medieval Medicus*.

gland, seemed ready for a technological breakthrough. Curiously, the evolution of the windmill from the water mill took a remarkably long time, even though most people were thoroughly familiar with the fundamental concept of using external force to generate rotary motion and the general level of technical expertise was fairly high. Earlier centuries had seen the development of numerous simple, but significant, improvements, such as the stirrup, the heavy plow, the horse collar, the horseshoe, and three-field crop rotation. One had only to look at the castles and churches rising everywhere—structures far more impressive than any that had been erected since the end of the Roman Empire—to realize that skilled carpenters and stone masons were working with new confidence. Military inventiveness had resulted in a redesigned knight's helmet, a more powerful crossbow, new stone-throwing catapults, and better siege tactics. A whipple tree had finally made harness hauling efficient, and oxen rather than horses were being used for bulk transport.[52]

The agricultural developments of the past few centuries were steadily being perfected and popularized. A spiked harrow was now used to smooth the soil, and an improved wheeled plow equipped with plowshare, colter, moldboard, and more iron components made tilling of the soil easier. More land was being devoted to legume crops, thereby increasing variety and nutrition in people's diets.

Large-scale engineering projects, such as bridge and road construction, were undertaken. Henry I had the Fosse Dyke recut to accommodate heavier water traffic, and the monks of Rievaulx Abbey redirected the course of the River Rye—an effort that

---

52. On technological and agricultural experimentation, see White, *Medieval Technology and Social Change*; *Medieval Religion and Technology*; and "The Expansion of Technology, 500–1500," in Carlo M. Cipolla, ed., *The Fontana Economic History of Europe*, 1:143–74. See also John Langdon, "Horse Hauling: A Revolution in Vehicle Transport in Twelfth- and Thirteenth-Century England"; and J. Z. Titow, *English Rural Society, 1200–1350*, esp. pp. 37–38. The literature in this field is growing and important.

required the construction of a series of artificial canals and took almost a century to complete.[53]

Perhaps most interesting, the Wigston Parva post-mill was not the first English effort to conquer the air. At least one attempt at manned flight preceded the perfection of wind-power technology. About A.D. 1000, more than a century before the arms of the first Western windmill fanned the sky, an adventurous monk from Malmesbury Abbey, in Wiltshire, had tried to fly.

Eilmer, as this eleventh-century aviator was called, lived a remarkably long time and enjoyed a wide range of experiences for a monk. He observed two visitations of Halley's comet, in 989 and 1066, and wrote a book on astrology that has since been lost. Malmesbury Abbey had long had a reputation for scholarship. One of its abbots, Saint Aldhelm (ruled c. 672–709), was a Latin scholar who wrote enigmatic poems on various topics, including the wind and even mills. He was also a noted architect and may have built the first pipe organ in Britain.

Eilmer conducted his experiment when he was probably less than twenty-five years old. He fastened wings to his hands and feet, hoping to soar like the fabled Daedalus. Then he climbed to the top of a tower, and when he felt a breeze, jumped off and flew for more than six hundred feet—a distance five times greater than that flown by the Wright Brothers at Kitty Hawk in 1903. His glide, or flapping transit, probably lasted for a couple of minutes. Eilmer became distressed because of the sudden violence of the wind, which made him aware of his own rashness. He plummeted to earth, breaking both his legs, and remained lame for the rest of his life. He is said to have ascribed his fall to having forgotten to put on a tail. This man was not a crackpot but a disciplined investigator who carefully measured the distance he flew and tried later to explain his crash.[54]

53. Trevor Rowley, *The Norman Heritage, 1066–1200*, pp. 70, 125.
54. William of Malmesbury, *G.R.*, 1:276–77. Lynn White, Jr., details this incident in "Eilmer of Malmesbury, An Eleventh-Century Aviator: A Case Study of Technological Innovation, Its Context and Tradition."

The available version of Eilmer's flight was recorded in approximately 1120 by another monk of Malmesbury Abbey, the historian named William. He had entered the abbey about 1090—too late to know Eilmer personally, but others there remembered him well. William's reputation and his contacts with important individuals, such as Bishop Roger of Salisbury, ensured that his works would be well received. Although there is apparently no connection between Eilmer's experience and the earliest post-mill, we can assume that as a result of William's account, alert readers in England in the 1120s were aware of the possibility of human flight.

This Malmesbury chronicler was not the only source of such information. Within a few years of William's account, Geoffrey of Monmouth, creator of much of the Arthurian material, depicted one of his characters, King Bladud, as being able to fly.[55] (Bladud was supposedly the founder of the city of Bath and the father of King Lear.)

William the Almoner moved in the circles in which such ideas about flying were examined. Whatever else they may have inspired, such discussions made it difficult for him to take the wind for granted. A few twelfth-century authors analyzed the wind, dividing it into numerous directional and descriptive types. However, their purpose was primarily devotional, for they wished to attribute spiritual significance, such as the presence of angels, to different types of wind. One practical thing that would have helped wind-power engineers was the kite; the Wright brothers used one extensively in their preflight experiments with air currents. Simple as it was, the kite was evidently unknown in the West until Marco Polo (c. 1250–1324) mentioned one in his journals.[56]

55. Geoffrey of Monmouth, *The History of the Kings of Britain*, p. 81.

56. For a description of a late-twelfth-century treatise on the "Various Significations of the Wind" from Llanthony Priory (Lambeth Palace MS. 392), see Montague R. James and Claude Jenkins, *A Descriptive Catalogue of the Manuscripts in the Library of Lambeth Palace*, p. 541. On Marco Polo's kite, see White, "Eilmer of Malmesbury," p. 105.

Angel or no angel, kite or no kite, when the dynamics of diverting wind to facilitate the age-old milling process were finally thought out, English carpenters made daring use of timber beams to coordinate the triple movements of billowing sails, grinding stones, and revolving buck. They also balanced very unequal forces and weights. However, there is no need to overintellectualize the origins of the windmill. After all, it was probably an unlettered person with calloused hands and a keen eye who put the first one together.

In addition to idealistic intentions, learned discussions, and technical skill, the concrete goals of inventors and promoters merit notice. This effort best focuses on farm management. Wigston Parva can serve as a typical microcosm of English rural life, for it was a manor, that is, a unit of agricultural exploitation—a farm, a fief, and an estate, as well as a village. It was leased by a tenant from a lord. Aluric and William the Almoner leased from the king; Arnold Wood, Gilbert, and Peter of Belgrave leased from the abbots of Reading.

The manor consisted of arable land (used largely for growing wheat, barley, and oats), some pasture and woodland, and a hamlet inhabited by the peasant villagers, free and serf, who worked the land. They tilled strips for themselves and demesne strips for their lord. The proportions of demesne and peasant acreage varied over the century, as did whether a lord managed his estate directly, or had a bailiff act for him, or leased the estate to a single renter (then confusingly called a farmer) for a fixed sum, or subdivided it. For example, after midcentury, Master Gilbert leased Wigston Parva from the monks for a set rental. The physician's son did likewise, but he allowed Ralph of Arraby to sublease the mill site. Thereafter, the abbey reasserted its control.

There were three main sources of manorial income: the rents paid by free and unfree peasants; the produce of the demesne, both grain and livestock; and miscellaneous fees, especially revenue from the local mill. Rents and mill profits could be fairly stable from year to year, but revenue from the sale of agricultural products fluctuated, depending upon weather and the health of animals and humans. Absolute, long-term increases in the income from these products might be expected only if more acreage were cultivated and if more grain were milled, that is, if there were larger harvests and more mills.[57]

Although perhaps not realized at the time, expansion was the watchword of the twelfth century. Population was increasing and towns were expanding. Europe was extending its internal frontiers, reclaiming forest, shore, and swampland at a furious pace.

New uses were found for barren land. Early in the century the Normans introduced fallow deer from the Near East and set them free to graze on acres too poor to farm. After their numbers increased they were hunted and became a regular part of some people's diets. Rabbits and pheasants were deliberately introduced into England for the same reason.[58] Systematic study of stock farming may have been encouraged at the large menagerie Henry I built for his own pleasure at Woodstock Palace.

Agronomy was not yet a serious study, witness the inadequate manuring of fields and the failure to rotate crops. However, times were changing. Henry I's clerks devised a set of questions about land management and annually checked the actions of royal bailiffs. Land reclamation was vigorously pursued and some farmers periodically fertilized their fields with lime. By 1200 agricultural productivity was being carefully studied, especially on large manors like those of the bishop of Winchester. Late-century crop yields, price patterns, and management techniques will be discussed in the last chapter.

A demonstrable relationship existed between managerial strategies and the state of the country's economy. Manors were often farmed out in economically troubled years, such as the pe-

57. Sidney Painter, *Studies in the History of the English Feudal Barony*, p. 153.
58. Rowley, *The Norman Heritage*, pp. 150–55.

riod 1139–1154. When stability returned, and particularly if prices rose (as they did late in the twelfth century), fixed rentals became unprofitable; consequently, landlords tried to repossess their leased lands. There was also a subtle, but definite, connection between land reclamation and the appearance of second-generation post-mills.

The people who began a technological revolution seem a more interesting subject than the economics of medieval farming. The ones to watch are the landlords, such as William the Almoner, who took a personal interest in their holdings; and the estate agents, such as Arnold Wood, who professionally supervised land development.

There has been a tendency to think of early barons as spoilers, or as absentee landlords. Many were not. Ernulf de Hesdin, one of the major beneficiaries of the Norman Conquest, is representative of the more forward-looking group.[59] By 1086 Ernulf held land in ten different counties directly from the king. A generous patron of several monasteries, he had once asked Geoffrey, Malmesbury Abbey's renowned physician, to cure his trembling hands. The doctor's ministrations were unsuccessful, but he suggested that the knight pray to Saint Aldhelm. It was balsam from his tomb that eventually cured Ernulf.

The baron was accused of treason against King William II in 1095. Not everyone so charged was actually guilty, however; Ernulf was vindicated in a bloody trial by battle, but he felt he had been disgraced. He thereafter became one of the few Normans to join the First Crusade and died at Antioch in 1098. His wife was still administering property according to his directions two years after his death.

According to a couple of monastic chroniclers who greatly admired him, Ernulf was an especially energetic lord who made

59. For a discussion of Ernulf, see Reginald Lennard, *Rural England, 1086–1135: A Study of Social and Agricultural Conditions*, pp. 50, 57, 64, 67, 69, 72, 85, 210–12; Frank Barlow, *William Rufus*, pp. 115, 279–80, 347, 358, 367.

many improvements on his estates. They considered him pious and businesslike and said he was intrigued by architecture and generous to the poor. Meticulous about tithing, he directed his bailiffs to ensure that his grain was properly assessed. Ernulf's success is evidenced by the increase in his total estate profits between 1066 and 1086. Although he has no known connection with the earliest post-mill, he does display the same combination of social concern and respect for manual labor possessed by William the Almoner. With people of this stamp about, it is no wonder that the fledgling post-mill spread its wings under Anglo-Norman skies.

# Spreading the News

---

*T*he first star of night comes out alone. It is followed by scattered pinpoints that shine in unpredictable places. Bright groupings then form meaningful constellations. Finally, broad highways of light traverse the skies. The early windmills can be compared to these glimmering pulsations. In the beginning they were few and far apart. Then they began to spread with mounting intensity. Each post-mill had special characteristics, but cumulatively, they would constitute a coherent enterprise.

From Leicester to London is a common trip, but the capital's windmill, possibly the second oldest in the land, is a star with a troublesome blinking quality. One moment it is plainly there. The next moment it seems to have disappeared. Its flickering identity is unique in the long register, yet it embodies numerous features possessed by the other mills.

In the late sixteenth century, John Stow, an avid antiquarian, compiled a survey of London. According to him, in 1100 Jordan Bricett established the Convent of Saint Mary of the Assumption at Clerkenwell and gave its chaplain and the Augustinian canonesses a site upon which to build a windmill.[1] Stow was definitely mistaken about the foundation date. The Clerkenwell nunnery was actually founded about 1144. The post-mill reference is harder to refute, or to substantiate.

Bricett's career permits no definite conclusions regarding his windmill sponsorship. A minor baron, younger son, and member of the country gentry, his patrimony was chiefly in Suffolk. His father, Brian, had also founded monasteries: an Augustinian priory at Bricett, on the family's ancestral lands, and a Cluniac priory at Stansgate in Essex. Jordan continued the practice. Also in 1144, just before founding Saint Mary's, he established at Clerkenwell a house of the new crusading order, the Knights of Saint John of Jerusalem. At about the same time he endowed a second, smaller house for the Hospitaller knights in Kent. Jordan's modest endowments for the Hospitallers and the Augustinian canonesses were his primary achievements of national importance, but whether he also helped the sisters build a windmill is a yet unanswered question.

The cartulary of Saint Mary's Convent at Clerkenwell preserves a copy of Jordan's donation. The charter clearly grants to the convent fourteen acres of land north of Saint John's Hospital and a site upon which to build a mill, but the cartulary copy specifies neither the precise location nor the type of facility.[2] One might assume that a water mill had been intended, but this was not John Stow's understanding, and he had once owned the cartulary. (It is he who wrote "1100" in the margin next to Jordan's proclamation.) Unfortunately, the cartulary copy cannot be checked against the original twelfth-century charter because the Clerkenwell convent's muniments were burned during the Peasant Revolt of 1381.

1. John Stow, *A Survey of the Cities of London and Westminster and the Borough of Southwark*, 2:85.
   2. *Clerkenwell Cartulary*, p. 30, charter 40; the charter is also printed in Dugdale, *Monasticon Anglicanum*, 6:807.

As indicated previously, scribes often used the terms "mill" and "windmill" interchangeably and indiscriminately. Moreover, Jordan, like many other post-mill enthusiasts (some of whom donated gifts to his two Clerkenwell foundations) gave the site for erecting a mill rather than a completed structure.

Clerkenwell was the site of a natural spring, but there does not seem to have been sufficient waterpower in the priory's field to make an effective traditional mill—at least that was the later opinion. Streams, and possibly mills, were located in other parts of Clerkenwell, however. Between 1170 and 1183, a gossipy clerk named William Fitz Stephen prefaced his life of Saint Thomas Becket with a short description of the city of London and observed that "there are on the north side pastures and pleasant meadow lands through which flow streams wherein the turning of mill wheels makes a cheerful sound."[3] Fitz Stephen never mentioned a windmill, but he omitted many facets of city life from this essay. He wrote in a rather florid, self-conscious style and concentrated on students, sports, and bustling street scenes.

An illustrated vellum roll of 1430 is more revealing. It includes a map of the estate of the London Charterhouse that shows that the Carthusians then operated a windmill, evidently near what would later be Windmill Inn in Saint John's Street. The map also reveals that a mound that had once been the site of a windmill had formerly been part of the Clerkenwell nuns' adjoining plot but that it was "now made plane with the field."[4] This leveled mound supports John Stow's contention that Jordan Bricett's grant to the canonesses included a windmill site. The neighboring Hospitallers also had a windmill. It cannot yet be dated earlier than 1338, but it is probably much older than that.[5]

3. William Fitz Stephen, *Life of Thomas Becket*, quoted in David C. Douglas and George W. Greenaway, eds., *English Historical Documents*, 2:957.

4. Bennett and Elton, *History of Cornmilling*, 2:251–52.

5. Lambert B. Larking and John M. Kemble, eds., *The Knights Hospitallers in England*, p. 94.

The study of windmills has occasionally been circumscribed by scholars' traditional beliefs. In 1185 a post-mill at Weedley, in East Riding, Yorkshire, was listed among the many possessions of the Knights Templar. It was then yielding an annual profit of eight shillings.[6] On the basis of this reference, some modern scholars have asserted that the Weedley windmill, and the new technology, were invented in 1185. Other commentators have observed that this particular installation could have functioned for many years before having been mentioned in the Templar survey. A careful analysis of the evidence suggests that this was indeed the case.

Three decades before 1185, a prominent local magnate, Roger de Mowbray, had included a promise to pay the tithes levied on unspecified mills at South Cave, Weedley, among his gifts to York Minster. He gave the hamlet itself to the Knights Templar.[7] This raises the possibility that the Weedley windmill(s) could have originally come from Roger and could have been at least thirty years old when the Templars tabulated their properties. Interestingly, Roger's gift to the cathedral had been in partial reparation for the damages he had caused during the reign of King Stephen.

Archaeological excavation may one day confirm an earlier date for construction of the Weedley windmill, but such a determination would be difficult because the village has completely disappeared. Aerial photographs do, however, reveal a plowed-out circular mound, undoubtedly the base of the vanished post-mill.[8] In the twelfth century, swamplands in this area of Yorkshire were being reclaimed, and some post-mills may have been used to drain water, as well as to grind grain. Because the Weedley windmill

6. Beatrice A. Lees, *Records of the Templars in the Twelfth Century: The Inquest of 1185*, p. 131; also see Appendix entry no. 56.

7. Diana E. Greenway, ed., *Charters of the Honour of Mowbray, 1107–1191*, p. 209, charter 324 (1154–1157). The Knights Templar also held land there belonging to Roger, some of which was granted to Saint Peter's Hospital before 1154; Saltman, *Theobald, Archbishop of Canterbury*, pp. 514–15, charter 285 (1150–1154).

8. Maurice Beresford and John G. Hurst, *Deserted Medieval Villages*, pp. 180–81.

was unusually profitable, it probably was used only for milling.

Another Yorkshire windmill might have been built even earlier. In 1190 a windmill at Beeford, in swampy Holderness, was given to the Cistercian abbey of Meaux. Three generations of lay owners are known to have owned it previously. It is not clear whether this means the mill was erected three years, three decades, or three score years before 1190.[9]

The number of post-mills was growing significantly by the 1160s. One sponsor was rather typical. William of Amundeville (d. 1168) was a knight who was the steward of the bishop of Lincoln. He and his family supported several religious institutions and also founded a hospital. William associated with physicians and knew other windmill enthusiasts. He was a generous, but not extravagant, benefactor of the Cistercians in Swineshead. He gave them just enough land on which to build a post-mill—precisely five strips.[10] Then, perhaps like Jordan Bricett, he allowed them to put up the actual structure themselves. William of Amundeville followed the more common practice of carefully distinguishing between the land beneath a post-mill, the facility itself, access to it, the profit from its production, and payment of the tithes based on that profit. Some lords even granted a miller along with a mill, but more often they donated the millman separately.

Tithes on the profits of windmill operations were considered very suitable gifts for charitable and religious institutions. In about 1170 Reginald Arsic made a grant of this type to the monks of Saint Mary of Hatfield Regis. His charter is particularly interesting because it still exists as an original document, with an appended white wax seal, rather than as an entry copied in a monastic cartulary. It was a chance reading of Reginald's statement, now preserved in the British Library, that convinced me that the 1185 date traditionally given for the invention of the windmill was in-

9. See Appendix entry no. 55.
10. See Appendix entry no. 20.

accurate. This accidental discovery was the first step in the long journey that eventually led to the writing of this book.

Let the benefactor once more announce his intentions:

> Reginald Arsic to all his men in the present and in the future, greetings. Know that I have given and granted and by this my charter have confirmed to the monks of Saint Mary of Hatfield Regis the whole tithe of my windmill which is in Silverley in the field which is called Breche in free and perpetual alms for the salvation of my soul and the souls of my father, my mother, and my ancestors. Wherefore I wish and direct that the afore-said monks hold the above-mentioned tithe happily and in peace from me and from my heirs.
>
> These are the witnesses:
> Ralf Arsic, my brother; William the Frenchman; Alexander Campium; Godfrey the dean; Ralf the clerk; Reginald the baker; William the mason; Ralf Musca; and many others.[11]

The declaration was written in a clear script, but the copyist made one striking error that he later erased. Initially he had written "windmills," but he changed the word to the singular form. This is a small point, but it nonetheless suggests the possibility that the scribe—who probably worked for the monastery rather than for the Arsic brothers—knew of more than one post-mill in Silverley. Ambition for his house evidently misdirected his pen as he momentarily thought that tithes from several mills would be coming to his priory.

Hatfield Regis, now called Hatfield Broad Oak, is a country village in Essex. About 1135 a priory of Benedictine monks was

---

11. B.L. Additional Charter 28340. See Figure 15. The author of the British Museum's old, handwritten, detailed *Catalog of Additional Charters* listed Reginald's grant as "late twelfth century." Apparently this description was based largely on paleography because the numerous other Hatfield Regis deeds are not related to this one in substance. The author of the printed catalog listed the grant as "temp Henry II," presumably on the same basis. The charter is printed without a date in H.M.C., *Seventh Report*, p. 589; and incompletely in G. Alan Lowndes, *History of the Priory at Hatfield Regis, Alias Hatfield Broad Oak*, p. 122.

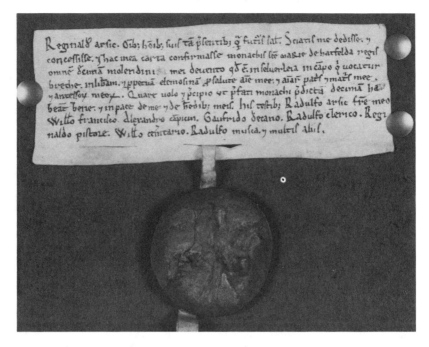

15. Reginald Arsic's original
charter granting tithes levied on his windmill at Silverley
to the monks of Saint Mary of Hatfield Regis.
Additional Charter 28340,
reproduced by permission of the British Library.

established there as a branch of the ancient Abbey of Saint Melanie, at Rennes in Brittany. The founder of the priory was Aubrey de Vere, who would later be the first earl of Oxford. Hatfield Regis Priory rose amidst the remarkable wave of monastic enthusiasm that swept Europe in the mid-twelfth century, but it never achieved particular spiritual renown or gained substantial wealth. However, an exceptional number of its original Latin charters, including that recording Reginald Arsic's grant, have weathered the ravages of time.

The hamlet of Silverley in Cambridgeshire was smaller than

Hatfield and is now considered part of the little village of Ashley, which is sometimes called Ashley-cum-Silverley. Silverley, like many settlements in the area, had a complicated tenurial history. Its parish church had been given to Hatfield Regis as an early endowment, but other local buildings, lands, and rents were divided among several knights and religious corporations.[12]

"Breche" was undoubtedly an ancient name for a field in Silverley—perhaps a recently cultivated meadow. However, the word might also have had a wider connotation. A couple of miles beyond the village, a large, vaguely defined region called the Brecklands stretches away to the northeast. This area was fairly prosperous from Neolithic to Saxon times, and such towns as Thetford achieved fleeting prominence. However, the sandy soil was only marginally productive, and the wind swept across it with unusual force.

This reliable air power attracted twelfth-century millwrights, who erected on the rolling Brecklands some of their best creations, such as the Elveden windmill. Rather careless, profit-driven ranchers also loosed great herds of sheep on the meadows. The unfenced animals voraciously grazed away the thin bracken, eventually trod the soil down to the chalk, and converted the area into a semidesert. Centuries later, the sheep would be better controlled and wide plots of pine would be replanted as windbreaks, enabling the Brecklands gradually to regain their present lush appearance. The record is vague, but conceivably, the Silverley windmill could have been the first mechanical invader of this fragile domain.

The Arsics were military tenants of the de Vere family, inconspicuously holding two hides (roughly 240 acres) from them in Silverley. In return for this small fief, they owed certain knightly

---

obligations, including a set period of military service. The relationship between a lord and his vassal usually extended beyond the responsibilities specified in contractual agreements because each conscientiously tried to promote the welfare of the other. Reginald must have felt this tug of chivalric fellowship when he decided to assist the monastery his lord had founded. However, his simple gift of windmill tithes also reflected major changes in world opinion.

Theoretically, a tenth of all the profits of all endeavors belonged to the church and should have gone directly to the nearest parish, to be used for its needs and for charitable distribution. Tithing is as old as Moses, but following the collapse of centralized authority during the decline of the Roman Empire and later in post-Carolingian Europe, powerful laymen appropriated these payments for their own use, just as they usurped the right to appoint bishops and abbots. Even minor barons regularly claimed multiple proprietary rights over nearby churches and their rectors.

Beginning in the mid-eleventh century, reformers in Rome and throughout the West courageously fought to abolish such lay domination of the clergy and to establish the supremacy of spiritual over temporal concerns. They also strove to increase the level of clerical education and to enforce clerical celibacy. The objectives of the so-called Gregorian Reform had been largely achieved by the mid-twelfth century. Nevertheless, some forbidden practices continued, such as the frequent lay appropriation of ecclesiastical tithes.

Continuing clerical opposition made peaceful lay exploitation of these prerogatives difficult, however. It seemed inevitable that these revenues would be restored to the church. Some landowners continued to dispute who should pay and who should receive mill tithes until well into the thirteenth century, but Reginald Arsic did not wish to become involved in such a controversy. His generosity to Hatfield Regis Priory and his willingness to cooperate with his own lord reinforced his pragmatic realization that the days when he could retain church tithes were numbered. Reginald undoubtedly recognized that he was in a tight situation and tried to make the best of it by donating to a monastery the tithes he had once collected and kept for himself.[13]

Properly speaking, such lay usurpations should have reverted to individual parish churches. Yet the papal reformers were so anxious to recover tithes and regain control over rectorships that monasteries were also permitted, even encouraged, to accept tithes and churches, on the theory that they could better defend them thereafter.

Some barons claimed that the windmill was an utterly new device and therefore was not subject to traditional tithing.[14] This rather specious argument was not advanced by Reginald. His initiative in voluntarily donating the tithe had allowed him to select Hatfield Regis Priory as its recipient. His foresight in this important matter calls to mind his thrift in continuing to use a parchment slip after a scribe had made a mistake on it.

Clearly, Reginald wanted all the local inhabitants to know what he was donating, for he called in as witnesses his own brother, the dean of the rural clergy, the baker, the mason, and many others. The absence of a miller is notable, but the pertinent question is, When did they all come together?

The scribe's handwriting suggests a twelfth-century date. Reginald's seal depicting him in knightly armor atop a galloping steed also reflects twelfth-century fashions, but it does not define the period very precisely. It is interesting that a minor landowner should possess such a handsome signet. The use of personal seals was for a time confined to kings and important churchmen, but

---

13. For a discussion of tithes, see Richard Mortimer, "Religious and Secular Motives for Some English Monastic Foundations"; and Giles Constable, *Monastic Tithes from Their Origins to the Twelfth Century*.

14. See Chapter 8.

the upper aristocracy subsequently adopted the practice. A justiciar once sneered that before 1165 only nobles were entitled to use private seals, but his recollection was incorrect. Physicians were using personal seals as early as 1110, and customs officials had employed them before that.[15]

Nevertheless, the justiciar was correct when he said that the practice was occasionally abused. To cite one instance, in 1182 there were thirty-three different seals being used at Bury Saint Edmunds.[16] No wonder that abbey was then deeply in debt. Reginald's seal indicates only that he had hired a good technical designer, that in using it he was not bashful about proclaiming his feudal status, and that his imagination far exceeded his modest means.

Reginald's predecessor in the Silverley fief was Geoffrey Arsic, undoubtedly Reginald's father, who can be traced to 1140 when he witnessed a charter of Aubrey de Vere.[17] In a comprehensive feudal survey of 1166, this Geoffrey was listed as holding two hides from Earl Aubrey.[18] Reginald succeeded sometime thereafter. Between 1188 and 1197, he accounted at the Exchequer for lands in Cambridgeshire and Norfolk.[19] He may have been ill

15. *Battle Chronicle*, p. 215.

16. Jocelin, *Chronicle*, p. 38.

17. B.L. Cotton Charter XXI.6, a grant by Abbot Ording of Bury Saint Edmunds to Aubrey de Vere I. It has been printed several times; see Douglas and Greenaway, *English Historical Documents*, 2:928.

18. *Red Book of the Exchequer*, 1:353.

19. *Pipe Rolls, 34 Henry II (1187–1188)* through *8 Richard I (1196–1197)*. Most entries through 1193 concern properties in Norfolk and Suffolk; subsequent entries pertain to Cambridgeshire. Both Reginald and Ralph Arsic attested a writ written by Brian Fitz Ralph for Bricett Priory; Cambridge University, King's College MS., Bricett Deed B 10. Reginald also witnessed a second deed (Bricett Deed B 114, group 1), as well as a grant of Aubrey de Vere I made about 1180; *Colne Cartulary*, p. 46, charter 88. Ralph Arsic attested for Aubrey's son before 1194. P.R.O. Duchy of Lancaster Miscellaneous Charter 249 (catalogued in P.R.O. unpublished D.L. *Cartae Miscellaneae*, 2:74, no. 249). The name of another Ralph appears between 1220 and 1230; see *Registrum Antiquissimum*, 4:22, charter 1136. Before 1194, the year of Aubrey I's death, Robert de la Mare gave to Hatfield Regis Priory a charter concerning land in Silverley next to the pond of Geoffrey Arsic (II); B.L. Additional Charter 28348. Presumably this

toward the end of his life, for by 1194 the Silverley tenement seems to have passed to his son, Geoffrey Arsic II, who was a minor at the time.

The most logical time for Reginald to offer a gift that would ensure the offering of prayers for the souls of his parents would be shortly after he assumed personal control of the fief—perhaps when he was still quite young. The striking absence of any reference to his wife or children suggests that he made the donation early in his career, before he married. Furthermore, one of the charter witnesses, Master Godfrey, seems to be the same man whose name appeared elsewhere as early as the 1130s. The names of some signatories, such as William the mason and Ralph the clerk, appear in other midcentury deeds for Hatfield Regis Priory. Thus, a date close to 1170 seems likely for this transaction.

Individuals bearing the surname Arsic (sometimes written Harsic) can be identified in several parts of the kingdom in the twelfth century, especially in Cambridgeshire, Lincolnshire, Norfolk, and Kent. In Oxfordshire, some Arsics were hereditary barons of Cogges.[20] Presumably all these families were related. Surnames were just coming into vogue in the twelfth century, and Arsic was one of the most distinctive. The Cambridge Arsics were evidently a comparatively unimportant branch of the clan, but like many richer families of the period, they exhibited a staunch conservatism in christening their children. For more than 150 years their baptismal names invariably alternated from generation to generation (Geoffrey, Reginald, Geoffrey, Reginald). Some Arsic descendants were also active windmill sponsors. Especially favored was the Hospital of Saint John the Baptist, founded in 1184 at Chippenham, not far from Silverley. Geoffrey Arsic III gave that hospital an annuity derived from a post-mill in Ashley,

---

Geoffrey was then still a minor and Robert was his guardian. Robert seems to have been in charge of Silverley in 1199 but apparently died the same year; William Farrer, *Feudal Cambridgeshire*, p. 247.

20. Sanders, *English Baronies*, p. 36.

which was next to Silverley.[21] Perhaps this was the same windmill the hasty scribe had had in mind decades earlier.

Windmill owners with business acumen, like the Arsics, soon learned that windmills, in addition to being useful to themselves, could also serve as impressive donations, and if the donation consisted only of the payment of tithes, it was relatively inexpensive. Such donors often favored recently founded religious communities, hospitals, and crusader preceptories. These institutions were pleased to receive the revenues, and they had the patience to wait for an eventual gift of the full installations.

Reginald associated with several other windmill enthusiasts, including the Bricetts, who endowed the convent at Clerkenwell. He also knew Master Stephen of Bury, son of Dean Herbert of Haberdun, who was the most outspoken proponent of the unrestricted use of air power. The prior of Hatfield Regis also knew the dean. Such links existed among many other clever, curious promoters. However, post-mills were essentially commercial operations. Altruism, and technological fascination, were quite minor priorities to most patrons. Reginald's donation of a tithe to Hatfield Regis Priory illustrates the attempt to combine generosity and self-interest; equally common, however, were fierce quarrels over the requirement that tithes be paid. Even more interesting was the belated seignorial effort to monopolize access to the new technology.

21. Before 1285, Geoffrey Arsic III paid the hospital an annual rent of thirty shillings; B.L. Cotton MS. Nero C IX, fol. 59. This Hospitaller cartulary includes numerous other gifts from Reginald Arsic's descendants. By 1338 the yearly contribution from the Ashley windmill had doubled, and the hospital had acquired two additional post-mills in Chippenham that were evidently somewhat smaller. The combined profits of these two mills were only sixty shillings a year; Larking and Kemble, *Knights Hospitallers*, pp. 121, 78.

# Enterprise and Restraint

"*E*rect no windmill or water mill to the detriment of existing mills."[1] Although not part of the Arsic declaration, such injunctions were common in many early title deeds. They highlight a serious dilemma that increasingly faced the owners of established mills. How could they encourage technological progress without losing control of their own monopolistic franchises?

Most of the thousands of operating water mills that were located throughout Norman England were owned by baronial magnates and religious institutions. The autonomous, individually owned and operated water mill was a rare phenomenon. Control over the milling process, or more precisely, the right to tax its profits, was traditionally considered a prerogative of the feudal aristocracy. Manorial proprietors therefore insisted that serfs grind their grain only at designated mills. Those same lords often tried to prevent the construction of milling facilities nearby so that free peasants and lesser knights would also have to use the baronial mill. Since most land was held contractually from some-

1. C.U.L. MS. Mm. 4.19 (Black Book of Bury), fol. 164.

one else (ultimately from the king), rather than possessed outright, such reservations could simply be added to the conditions of tenure. Moreover, it was easy to control water mill proliferation by limiting access to the best sources of water power.

The innovative windmill upset this careful balance. Coincidentally, it appeared just as the growing population's need for ground meal was beginning to exceed the productive capacity of the old water mills. Some farsighted landlords anticipated that the new apparatus might offer a solution to this problem and erected additional mills. In a sense, this expansion violated the baronage's own philosophy of limited installations, but there never had been a uniform policy.

To the distress of all mill proprietors, certain adventurous upstarts, who realized that the wind was freely available to everyone, learned that post-mill machinery was relatively inexpensive and that there was a genuine demand for more processed flour. These brash independents—clever peasants, knights dissatisfied with their small fiefs, university-trained intellectuals, and women anxious to support themselves—soon built their own rival mills.

Baronial reaction was immediate and intolerant. Determined to safeguard hereditary monopolies and to preserve the social hierarchy, many concerned magnates prohibited the construction of any windmill or water mill that they did not themselves license. The resulting entrepreneurial struggle can be observed from the viewpoints of both the established authorities and the defiant newcomers. Four generations of an influential baronial house, the Clares, illustrate the pride and prerogatives of the incumbent mill owners. One elderly rural dean epitomizes the aspirations of the aggressive challengers (Chapter 6).

The Clares were major beneficiaries of the Norman Conquest. Even the cadet branch, which is important for this study, was one of the wealthiest families in England. For much of the Anglo-Norman period its leader was Walter Fitz Robert (c. 1126–1198), a tough, articulate warrior and an advocate of wind-power technology and monopolistic control of milling. His parents, his children, and his grandchildren were also extraordinarily attentive to mill franchises. Walter and his relatives promoted at least six post-mills and adamantly refused to allow anyone to construct competing facilities. The baron's order, "Erect no windmills or watermills to the detriment of existing mills," sounds like the commercial battle cry of his whole social class.

Walter's father was Robert Fitz Richard of Clare, master of Baynard's Castle in London, lord of the barony of Dunmow in Essex, and hereditary steward of the kings of England. Dozens of knightly vassals owed him military service and he received revenues from numerous estates. Walter's mother was Matilda of Senliz, the daughter of the earl of Huntingdon; her mother later married King David of Scotland. Matilda was a domineering person, and she, even more than her husband, determined the patterns Walter would follow all his life.

Baron Robert died about 1136, evidently when Walter was still a youth, perhaps about ten years old. By 1146 his mother felt that she could marry again and that Walter could manage his paternal inheritance directly. One of his first tasks occurred between 1148 and 1153 when he used the royal courts to recover some ancestral lands that had been illegally occupied by local troublemakers. Before 1154 he received a writ from King Stephen ordering him to safeguard some property for Colchester Abbey, a foundation of his aunt, but he took no notable part in the civil war of Stephen's reign.

Matilda of Senliz married Saher de Quency I (d. c. 1158) and began a second family. She never neglected her eldest son and Walter cherished his half brothers and sisters. One of them, Saher de Quency II (d. 1190), wed an heiress of the Bourn estate in Cambridgeshire. It would be fitting if he shared the post-mill enthusiasm of his mother and brother. Perhaps Saher and his wife erected the ancestor of the present Bourn windmill, and perhaps that sponsorship gave rise to the confused tradition that Sheriff Picot built a post-mill in Cambridge a century earlier.

Walter Fitz Robert's baronial career spanned more than six

decades, an impressive tenure in any era. Family concerns were a principal focus of his life. He maintained the economic and philanthropic policies of his parents and inspired his children to preserve them as well. He made two brilliant marriages, first with Matilda de Lucy, daughter of Richard de Lucy, the justiciar. After her death in the early 1160s, he wed Matilda de Bohun, a widow and the daughter of the earl of Hereford. Walter had at least eleven sons and two daughters in his long life. He was not entirely faithful to his wives, however, for at least one son was illegitimate. The matter was evidently smoothed over, for Walter acknowledged him and yet the lad remained proud of his mother. The baron also suffered the heartache of having two of his sons, including his principal heir, Richard, die before him. As if to compensate, Walter seems to have become increasingly fond of a son who was a priest.[2]

Walter's public career was eventful, but not as truly distinguished as might have been expected from so powerful a person. He had estates scattered throughout six counties, from Kent to Northamptonshire, but most were in East Anglia. His barony entailed more than sixty-six knights, almost one percent of the kingdom's entire complement. He inspired unusual, lasting affection among those retainers, many of whom piously mentioned him in their donations to religious houses.

Walter's royal stewardship was largely ceremonial, but he was unfailingly loyal to whoever was king. This was of considerable consequence because his control of Baynard's Castle probably

2. Baron Walter's wives were badly confused in Dugdale, *Monasticon Anglicanum*, 5:147. No thorough study of his life exists, but see Sanders, *English Baronies*, p. 129 (which lists his wives in reverse order). For the later Clares see Michael Altschul, *A Baronial Family in Medieval England: The Clares, 1217–1314*. For Walter's earliest appearances see *Regesta*, vol. 3, charter 239c; Helen Cam, "An East Anglian Shire-moot of Stephen's Reign, 1148–1154." His illegitimate son was Humphrey, the son of Agnes; B.L. Harley MS. 662 (Dunmow Priory Cartulary), fol. 63v. Richard attested two charters as Walter's heir; B.L. Cotton MS. Claudius D XII (Daventry Priory Cartulary), fol. 6 (after 1164); and *Knights' Cartulary*, p. 202 (about 1170).

meant he also commanded the London militia. He certainly served Henry II with great enthusiasm—as a military officer and as a diplomat in Normandy and on the Welsh frontier. For decades he also served as an itinerant justice, periodically hearing pleas in eight shires almost until the year he died.

The steward suffered some sort of personal crisis about 1170, perhaps a serious illness compounded by the deaths of his sons. He put his affairs in order and made plans to dispose of his warhorse and armor, but he recovered and flourished for more than a quarter of a century, until 1198.

Some of his most exciting moments came in October 1173, during the rebellion of Henry's son, Henry, the so-called Young King. Walter and other loyalists were stationed at Fornham, in the marsh near Bury Saint Edmunds, when they suddenly found themselves badly outnumbered, perhaps four to one, by the invading army. The raiders were Flemish allies of the rebels, and they boasted that they had come to destroy the old king and to seize his wool. Although well into middle age at the time, Walter Fitz Robert impetuously attacked the enemy head on. His first charge was repulsed by the Flemings, many of whom were really weavers, not soldiers, but he rallied and eventually routed the cloth troops. Years later, in 1191, the baron defended the rights of another sovereign, the absent crusader king Richard the Lionheart. This time he foiled one of the schemes of Prince John; in reward he gained the constableship of Eye Castle in Suffolk.[3]

Despite such adventurous national service, Walter's major influence was in local affairs. In that respect he resembled most of the other windmill pioneers. He knew promoters like the Arsics and the Bricetts, but was far above them on the Anglo-Norman

3. For Walter's crisis see notes 20 and 21 to this chapter. For his Welsh activities see Giraldus Cambrensis, *Opera*, 1:58–59. For Prince John's troubles see Harry Rothwell, ed., *English Historical Documents*, 3:66–67. For the battle of 1173 see Jordan Fantosme, *Chronique de la guerre entre les Anglois et les Ecossais en 1173 et 1174*, pp. 289–91.

social scale. However, like them, he realized that mills and mill revenues were particularly appropriate gifts for religious institutions. His charters record such endowments time and time again. Many of his writs were unusually detailed, and eighty-three of them have been recovered—a staggering total for a layman of this period. The baron apparently maintained a private chancery for such secretarial needs. Several of his documents were attested by Edmund the clerk, and another clerk named Hamelin noted that he drew up some of the charters.

Administrative expertise also characterized Walter's management of his estates. He was far more aware of economic inflation than were many of his peers. He was well informed about the most minute workings of his manors, had a special interest in grain milling, and frequently asked millers to attest his charters. He also knew a great deal about the religious houses he championed.

Walter endowed at least fourteen institutions. He most favored Daventry Priory, a Northamptonshire monastery dear to his mother, and Dunmow Priory, an Essex house patronized by his father. In addition, he aided Bricett Priory in Suffolk, Saint John's Priory in Clerkenwell, Bury Saint Edmunds Abbey in Suffolk, Saint Bartholomew's Hospital outside London, Colchester Abbey in Essex, Saint Neot's Priory in Huntingdon, Dereham Abbey in Norfolk, Bruern Abbey in Oxfordshire, Stoke-by-Clare Priory in Suffolk, Saint Andrew's Priory in Northamptonshire, Thorney Abbey in Cambridgeshire, and an unnamed hospital in Dunmow. Except for his probable creation of the Dunmow hospital, he did not found any institutions himself. Rather, he was content to enhance the work of others, an unusually modest trait for someone of his stature.

Walter Fitz Robert was a vigorous competitor in war and in business. He reacted quickly to opposition and only foolhardy individuals crossed him twice. Otherwise, he seems to have been a thoroughly likable person, a man who embodied the values thinking people always treasure—family love, religious devotion, neighborly concern, personal courage, and technical progress. In 1198 he was succeeded by his eldest surviving son, Robert Fitz Walter, who became the dynasty's most famous member, primarily because of his successful opposition to King John during the struggle for Magna Carta. Robert was an arrogant man with an ironic sense of humor, but he shared the family fascination with all aspects of milling.

Walter's clan donated mill rents, mill sites, individual mills, tithes levied on mill profits, and even millers to favored monasteries and personal dependents. For example, Walter's parents gave a water mill to Thorney Abbey, and his mother granted three mills to Daventry Priory. Walter later ratified these gifts and added endowments of his own.[4] Eventually, he distributed rights in more than a dozen different mills. Surprisingly, his gifts were usually derived from a very limited number of estates. Thus, he granted Saint Bartholomew's Hospital an annual stipend of five shillings from the mill (unspecified) in his park at Woodham in Essex.[5] He bestowed the right of nominating the rector (the advowson) of Woodham Church on the Hospitallers of Clerkenwell.[6] They also received four shillings annually from the Woodham mill.[7] The manors of Burnham and Dunmow in Essex, Daventry in Northamptonshire, and especially Hempnall in Norfolk were other sources of multiple benefactions. Such generosity was not altogether altruistic. Tithes, advowsons, and other lay perquisites were being reclaimed by reforming churchmen. Like

4. C.U.L. MS. 3021 (Thorney Abbey Cartulary), fol. 323; B.L. Cotton MS. Claudius D XII, fols. 1, 5v; Dugdale, *Monasticon Anglicanum*, 5:181.

5. *Saint Bartholomew's Cartulary*, p. 121, charter 1260 (about 1190). This charter mentions both of Walter's wives.

6. B.L. Cotton MS. Nero E VI (Hospitaller Cartulary), fol. 124 (1165–1170); printed in Dugdale, *Monasticon Anglicanum*, 6:807. This grant was confirmed by Bishop Gilbert Foliot between 1165 and 1174; Adrian Morey and Christopher N. L. Brooke, *The Letters and Charters of Gilbert Foliot*, p. 428, charter 385.

7. B.L. Cotton MS. Nero E VI, fol. 124. Walter made other grants to the Hospitallers of Clerkenwell, including land near a holding of William the miller (fol. 102v).

several clever magnates, Walter was systematically divesting himself of controversial holdings while proclaiming he was making voluntary contributions.[8]

Little Dunmow, a small village in Essex, was the Fitz Robert baronial seat, and the Augustinian priory Walter's father had established there was a special object of Clare pride. It was the family burial place; the bodies of Walter and other members of his dynasty still rest in its chapel. His favored son, Walter Fitz Walter, became a canon there, apparently some years after having served as a secular clerk. This was a refreshingly humble vocation for the son of such a prominent baron, but the lad was amply supported because he had both feudal and ecclesiastical preferments.

Land he held in tenure from his father was given to the priory when he became a canon. Moreover, in 1184, the baron gave Dunmow Priory the Church of Burnham in Essex; young Walter was subsequently made vicar there.[9] The elder Walter also confirmed the priory's tithes of the mills in Dunmow and, even more generously, granted his entire mill in Burnham, which may have been a tidal mill, since it was described as being located in salt water.[10] In characteristically complex fashion, the knight further offered the Hospitallers at Clerkenwell advowson rights with regard to Burnham.[11] Walter Fitz Walter had yet another benefice,

8. For a discussion of this policy, see J. C. Ward, "Fashions in Monastic Endowment: The Foundations of the Clare Family, 1066–1314."

9. B.L. Harley MS. 662, fols. 7v, 11v, 68v–69. Prior Ralph of Dunmow and Archbishop Richard of Canterbury confirmed Walter's vicarage between 1181 and 1184; *Hospitaller Cartulary*, pp. 313–14. There is some confusion about the holdings at Burnham. The editor of the Kentish portion of the *Hospitaller Cartulary* thinks the parsonage was in Kent, near the bank of the Medway; *Kentish Cartulary*, pp. 3, 134.

10. B.L. Harley MS. 662, fols. 6v, 69. The Dunmow tithes had actually been given earlier by Walter's father. Bishop Gilbert confirmed the grants of Walter and his father between 1167 and 1180; Morey and Brooke, *Gilbert Foliot*, p. 361. The Burnham mill may have been located on the River Crouch at the point where the clear water meets the salt water.

11. B.L. Cotton MS. Nero E VI, fol. 124 (1165–1170); printed in Dugdale, *Monasticon Anglicanum*, 6:807.

the prebend of Finsbury in Saint Paul's Cathedral, London.[12] The cathedral was extremely close to his father's tower at Baynard's Castle. The son never rose above the modest rank of cantor in the chapter, but one cathedral property he held in Hertfordshire may have supported an early windmill.[13]

Some of the royal steward's vassals also enriched Dunmow Priory, but he was the principal benefactor. On its behalf, Lord Walter promulgated forty-one different charters, approximately half of his corpus. Many charters recorded gifts of parishes and lands, such as the Church of Hempnall and the tenement Eudo the physician once held in that manor.[14] On the whole, the few restrictions on his gifts were set by his mother's example. His father had routinely given the Dunmow canons tithes of all the mills in their village, but when Matilda of Senliz later decided that a local priest could build his own mill in Dunmow, she made him first offer the canons twenty shillings for the privilege.[15] Her intent was crystal clear: always protect the standing facilities.

In succeeding centuries, local residents would remember the family for reasons quite apart from its mill gifts. According to one hardy legend, Walter's heir, Robert Fitz Walter, resolved late in life to refurbish the priory and to increase its endowment. He created a special fund to be administered by the canons for a rather comical ironic purpose. Robert declared that any man who did

---

12. Diana E. Greenway, *John Le Neve's Fasti Ecclesiae Anglicanae, 1066–1300*, 1:23, 49. He was a canon about 1184, cantor, or precentor, after 1189, and died in 1203. Since it would be unusual for an Augustinian canon to hold a benefice in a secular cathedral, there may have been two Walters Fitz Walter. Furthermore, the Walter at Dunmow seems to have died before 1198; B.L. Harley MS. 662, fol. 68v. If there were two young Walters, perhaps one was another illegitimate son of the baron.

13. The evidence is only suggestive, not conclusive. In 1222 there was a windmill (*molendinum ad ventum*) on Saint Paul's manor, Caddington, in Hertfordshire. William the treasurer held that manor as his prebend from 1192 to 1222, but at some point Walter Fitz Walter held land there. The post-mill was described as needing repairs, so it may have been quite old in 1222; *Domesday of Saint Paul's*, pp. 1–2.

14. B.L. Harley MS. 662, fols. 7v, 94.

15. Ibid., fol. 11.

not regret his marriage, while either sleeping or waking, for a year and a day might lawfully come to Dunmow and claim a ration of bacon. Before the prize could be awarded, the pilgrim had to kneel on two sharp stones in the priory courtyard and swear that he was continuously happy. Then the canons chanted at length; the townspeople carried the hero around on their shoulders and, finally, the husband literally brought home the bacon.[16]

If Baron Walter's son did in fact inaugurate this engaging custom, he may have intended it as a wry commentary on life. His father, his stepmother, his grandmother, and his great-grandmother had all been married twice. Appropriately, the first literary reference to the "Dunmow Flitch" is that of Chaucer's garrulous, much-married Wife of Bath.

Although Dunmow Priory was the special object of Clare philanthropy, Walter Fitz Robert gave one of his most expensive mill gifts to the mighty Abbey of Bury Saint Edmunds, perhaps to commemorate his nearby military victory. Bury employed resident physicians, financially supported several local hospitals, encouraged a succession of superb manuscript illuminators and historians, used sophisticated plumbing and a water clock, and profited from the construction of multiple windmills. It also incurred seemingly endless debts.

Walter's feudal relationship to the abbey was relatively minor—merely a few isolated properties and a small fief held of the monks for the service of one knight.[17] His contributions were truly substantial: a miller named William and his tenement, a whole water mill, and the land upon which to erect a windmill. Although they came in three separate grants, all these endowments were based on Walter's small manor of Hempnall in

16. Dugdale, *Monasticon Anglicanum*, 6:149.
17. Samson, *Kalendar*, p. 66; also see pp. 25, 27, 32, 69, for mention of his isolated small holdings.

Norfolk, a manor over which the abbey claimed ultimate jurisdiction.[18]

William the miller might have been a serf, for he was traded like a chattel. However, the grant probably entailed only his services, not his person. In prosperous Norfolk, which boasted an unusually high percentage of free peasants, this miller was most likely a free manorial tenant. Locally, William was a person of consequence, for Sir Walter asked him to attest some writs. The baron would hardly have called upon a serf to pledge himself in such matters.

Twelfth-century Englishmen lovingly personalized their machines, just as their descendants had done. The Hempnall water mill rejoiced in the apt, homey name Twigrind.[19] Walter's gift prescribed the whole mill, except for certain fishing rights in its pond. He announced that his donation was made on behalf of the souls of his parents and his first wife, Matilda de Lucy, as well as for the salvation of his second wife, Matilda de Bohun, for his heir, Robert, and for all his successors. He directed that at his death, the Bury monks were to enroll him in their perpetual confraternity of prayers and masses. Finally, he stipulated that the monks were to be given one silver mark each year when they

18. The charters recording the Hempnall grants were copied in several Bury cartularies: C.U.L. MS. Mm. 4.19 (Black Book of Bury), fols. 163v–64v; C.U.L. MS. Ff. 2.33 (Bury Sacristan's Register), fols. 55v–56v. The windmill and water mill grants were mentioned in B.L. Additional MS. 14847 (White Book of Bury), fols. 22–22v, but the full text was not copied by the scribe. On the jurisdiction of Bury Saint Edmunds over Hempnall, see Cam, "An East Anglian Shire-moot."

19. The name Twigrind appeared in another Bury Saint Edmunds document as early as 1145; David C. Douglas, *Feudal Documents from the Abbey of Bury St. Edmunds*, p. 159, document no. 180. However, this charter might refer to Cavenham, in Suffolk, not far from Bury where the Clare earls of Hertford had mills called Twigrind, a portion of whose rents they gave to the monastery of Stoke-by-Clare; *Stoke-by-Clare Cartulary*, 1:25, charter 37 (1152–1173), and p. 59, charter 71 (1152–1173). Sometime later, the whole Cavenham estate passed to the Knights Templar, who called it Togrind. Eventually this property would be administered by the nearby Templar preceptory of Ashley-Silverley in Cambridgeshire; Dugdale, *Monasticon Anglicanum*, 6:833.

celebrated the feast of the glorious martyr, Archbishop Thomas, and another when they commemorated the anniversary of his own death. If the profits of the water mill increased in the future, the brothers should receive even more. This unusual provision, a veritable escalator clause, suggests that Walter was well aware of the effects of inflation.

The reference to Thomas Becket means that the Twigrind donation was made after 1170, when the archbishop was murdered, and probably after 1173, when he was canonized. It is surprising to find Walter, an ardent king's man, advocating devotion to the champion of antimonarchical resistance, but he even named one of his sons Thomas. The cult of the martyred archbishop spread like wildfire and attracted all sorts of devotees throughout Western Europe. Even King Henry II did penance and made his peace with the saint.

Walter seems to have been momentarily preoccupied with death. Perhaps he was gravely ill. Besides securing spiritual remembrance by means of his Twigrind grant, the baron, about 1170, directed that at his death, his warhorse and armor were to pass to the Knights of Saint John.[20] He managed to have his name listed in other monastic memorial rolls, including that of distant Durham Cathedral priory.[21] Since Robert Fitz Walter was specified as his father's heir in the Twigrind grant, the baron's eldest son, Richard, must have already died.

Whether he was under stress, or not, the lord of Dunmow never lost sight of his commercial objectives. In addition to the Twigrind water mill he had granted to Bury Saint Edmunds, Walter and his heirs sent the monks a handsome annual cash payment. Obviously, Walter intended to continue to supervise this profitable business personally. The Bury monks undoubtedly found this a convenient arrangement, since the workshop of the

---

20. B.L. Cotton MS. Nero E VI, fol. 335; printed in *Knights' Cartulary*, p. 202, charter 344.

21. *Liber Vitae Dunelmensis; nec non obituaria duo eiusdem ecclesiae*, p. 19.

miller (who was now their own man) was a considerable distance from their abbey.

Walter's grant of the Hempnall windmill site was also carefully detailed. He reported giving the monks of Bury Saint Edmunds one-half an acre at a place called Longbridge on which to erect a post-mill. This was a plot of rather modest dimensions, but the land was probably particularly valuable. Apparently it had a special feature—perhaps an unusual elevation above sea level. It was in the eastern part of the land belonging to an Adam of Hempnall. In order to obtain it, Walter had exchanged with Adam half an acre in an area called Harlesgappe (?). One-half acre was about the size of the plots of some other windmill pioneers, but the baron probably also paid for construction of the post-mill. Local opinion consistently attributed it to his generosity.

The monks were to enjoy the same liberties with respect to the windmill that they enjoyed with the Twigrind water mill. Walter vigorously commanded that neither he, nor his heirs, nor his retainers, were to erect in Hempnall any additional windmill or water mill to the detriment of the monks of Saint Edmund. This provision was a key component in the charters recording other windmill grants as well. In addition, the steward directed that no Hempnall resident was to go to another mill to grind grain if the grinding could possibly be done at Longbridge. Clearly, Walter was restricting competition in Hempnall and forcing people there to use certain mills.

Since Walter mentioned the Twigrind water mill in his grant of land on which to build a post-mill, the Longbridge windmill must have been erected after Bury's acceptance of Twigrind. If the water mill was donated to the monastery only after 1170–1173, as seems likely, the post-mill would necessarily have been built after that. It is puzzling, however, that Walter's windmill declaration referred only to one unnamed wife, whereas most of his title deeds after 1164 included specific allusions to both of his ladies. Thus, the Hempnall windmill is truly datable only to

1136–1198, the span of Walter's career. Shortly after the baron's death, his three grants of William the miller, the Twigrind water mill, and the Longbridge windmill were confirmed by his son, Robert, in a brief writ.[22]

Walter's second bride had some windmill associations of her own. Matilda de Bohun (her maiden name) was also called Matilda d'Oilli, after her first husband, Henry d'Oilli I (d. 1163), a royal constable. Their son, Henry d'Oilli II (d. 1196), sponsored construction of a windmill at Cleydon in north Oxfordshire and permitted the monks of Oseney Abbey free access across his land to the mill. Young Henry witnessed a writ of Walter's and Matilda's son, Simon, aimed at preventing competition with the mills of Daventry Priory.[23]

Walter Fitz Robert's bridal gift to Matilda de Bohun included the manor of Henham in Essex, not far from Dunmow. Approximately four decades later, in 1202/3, a windmill at Henham was casually mentioned in a lawsuit when Matilda, then Walter's widow, was reconfirming her other rights with regard to the manor. That post-mill had undoubtedly been operating for several years. Thus, in addition to it, Matilda could claim windmill connections via her son by her first marriage as well as her second husband. Walter's successor, Robert, also married a woman with windmill connections. Gunnora was the sister of Theobald of Valeins, the principal post-mill proprietor along the Norfolk coast.

Clare family interests extended far beyond Essex, London, and Norfolk. The quaint country village of Daventry, in Northamptonshire (less than ten miles from Cleydon) was particularly important to the Clares. That settlement was in the heart of the Midlands, in the shadow of one of England's largest prehistoric

22. See footnote 18 to this chapter.
23. On the mill, see Appendix entry no. 41. On Henry's attestation, see B.L. Cotton MS. Claudius D XII, fol. 6v.

hillforts. About 1107/8 Walter's maternal grandfather, Simon of Senliz, earl of Northampton, assisted a small group of Cluniac monks to leave their unsuitable site at Preston Capes and take up residence in Daventry. To the small priory's endowment he then added gifts such as tithes of the Daventry mills. A hospital was built in Daventry sometime thereafter.[24]

Simon's daughter, Matilda of Senliz, became very attached to the monastery, and perhaps for that reason, her father made the manor of Daventry part of her dowry. These lands remained under her control throughout her two marriages and widowhood, and she continuously supported the monks. Matilda's encouragement took many forms, but the monks especially welcomed her gift of mills. The brothers were intensely interested in milling, and they accepted and built facilities far beyond their needs. By 1200 they had benefited from at least one dozen installations in the immediate area and on nearby estates. The post-mill component of this massed power is seemingly mired in contradiction. Even the scribes who prepared the abbey cartulary could not keep all of the Clare donations straight.

After her second husband's death in 1158, Matilda of Senliz gave the Daventry monks three unspecified mills and pointedly forbade anyone to erect competing mills in Daventry or in nearby Drayton, except for the monks' own benefit.[25] One of the mills was a windmill, but the donor felt no pressing need to specify it as such. Three succeeding generations would issue similar generalized declarations. When Walter confirmed Matilda's grant before 1166, he added a few minor gifts of his own and in his grant

---

24. For a general history, see *V.C.H., Northamptonshire*, 2, pt. 1, p. 110; Dugdale, *Monasticon Anglicanum*, 5:178–81. The hospital is not listed in David Knowles and R. Neville Hadcock, *Medieval Religious Houses: England and Wales*.

25. B.L. Cotton MS. Claudius D XII, fol. 5; Dugdale, *Monasticon Anglicanum*, 5:178–81. The last mention of Matilda seems to be in 1158, but she did live for a while after her second husband's death.

mentioned John the miller, who was evidently operating the machines.[26] The baron did not specify the mills, as his mother had not. Probably during his second marriage, Walter reissued his confirmation, perhaps anxious that children from both his marriages should clearly understand his instructions.[27]

Walter had a strange encounter with one of his Daventry vassals relating to yet another mill. In effect, this man was testing the authority of Matilda and Walter. According to the baron's account, a William of Drayton erected a mill, also unspecified, next to his own house. Walter obviously considered this a violation of his express prohibition, and he counterattacked as soon as he could give the matter his direct attention. In a compromise settlement, he required William to surrender the offending mill to him as the overlord. Then, instead of demolishing it, as he might well have, he gave it to Daventry Priory.[28]

The terms of the agreement were strict and unusual. Walter does not seem to have wanted anything for himself except the satisfaction of having imposed his will. Following William's capitulation, the Daventry monks were to lease the Drayton mill back to him for the yearly rental of twelve pence. This was a modest sum, but then the knight could expect no income from the mill. He could use it for his own grain, but no one else's, not even the Daventry monks. For some unknown reason, Baron Walter further insisted that only water from Saint Etheldreda's well could be used at the mill. William of Drayton and his son, Hugh, had to swear to all these conditions "on the church's book," and they were warned in the charter that the mill would be razed if they broke the covenant. It is hard to understand how this last

26. B.L. Cotton MS. Claudius D XII, fol. 5. Walter also gave the Daventry monks fishing rights, permission to build a road, timber, and herrings.
27. Ibid., fol. 6.
28. Ibid., fol. 6v. The original charter recording the donation still exists in a collection of Daventry charters in the Oxford University Bodleian Library, where it is cited as D.7. It is summarized in *Christ Church Cartulary*, pp. 53–54.

provision could have been enforced, considering that, according to the agreement, the mill belonged to the monks, but Baron Walter was concerned about competition, not legal niceties.

Yet, for all Walter's bluster, and considering the gravity of William of Drayton's direct challenge of seignorial authority, Walter's charter was still a compact between chivalric gentlemen—between a lord and his vassal—not between enemies. William clearly lost, but not much, unless he had been planning some dramatic milling expansion on his own. He was not humiliated and his superior's order had been upheld. Moreover, both laymen had the consolation that they were benefiting the church.

The provision that only Saint Etheldreda's water be used seems curious. Except to quench employee thirst and for fire prevention, it is unclear why any type of mill would need well water. Surely a water-powered mill had no additional need for well water, but a windmill did not use much water, either. Milling flour was essentially a dry process. This reference to the saint's spring does make slightly better sense, if William of Drayton's installation was a windmill. Happily, later evidence confirms this probability. The mill, which was subsequently called Ryen Hill Mill, was located in the south field of Drayton and was indeed a post-mill.[29]

There were thirty witnesses to the surrender of the mill—an indication of the punishment's importance and severity. Clearly, a lesson in feudal prerogatives was being publicized. Leading the long list of signatories were the names of Walter's second wife, signing as Matilda de Bohun, and of her distinguished sister, Margaret de Bohun. This transaction, and the construction of the Drayton windmill, can confidently be dated as occurring between 1164, the earliest possible date for Matilda's marriage to Walter Fitz Robert, and 1197, the year of Margaret's death.

The final witness was William Gemme, who added a postscript after his signature, in which he declared that it was he who

---

29. Ibid., pp. 56–57. See also note 35 to this chapter.

surrendered the mill into his lord's hands before Walter gave it to
Daventry Priory and that he ought not to have erected or held a
mill "against the liberty of" Daventry. It would seem as though
William of Drayton, the builder of the mill, and William
Gemme, who said he erected and surrendered it, were one and
the same. If so, he does not appear to have defended his action. In
a remarkably similar confrontation at Bury Saint Edmunds, the
landlord would endure much stronger opposition (see Chapter 6).

If William Gemme were someone else entirely, his role in
this controversy would appear to have been that of a valued me-
diator. It was not unusual for a twelfth-century person to use more
than one name, such as William of Drayton and William Gemme
of Daventry. Remember that Walter's second wife was known as
both Matilda d'Oilli and Matilda de Bohun.

William Gemme of Daventry was certainly known to Baron
Walter, who about this time gave him a small parcel of land in
Daventry.[30] Whether William was already in the royal steward's
debt when the windmill dispute occurred, or whether this was
some sort of subsequent compensation, cannot be determined.
William of Drayton was, of course, Walter's vassal.

Interestingly, there was yet another William connected with
Daventry at this time: William the miller, who leased the mill
(unspecified) of Cudon from Prior Reiner (c. 1170–1190) for a
rent of twenty shillings.[31] The witnesses to the agreement included three other millers. Perhaps this William was the same
miller involved with the Hempnall post-mill and (or) the same
man who attested some of Baron Walter's writs. If so, he emerges
as a type of miller-in-chief. This is quite plausible, since in the

---

30. *Christ Church Cartulary*, p. 57; see also pp. 99, 100, 119. B.L. Cotton MS.
Claudius D XII, fol. 54v, includes a cartulary copy of the charter recording Walter's
gift to Walter Gemme. The names of some of the witnesses also appeared elsewhere in
the 1160s.

31. *Christ Church Cartulary*, p. 54.

next generation a superintendent miller named Roger was an important local figure in Daventry and Drayton, and a similar position seems to have existed in Holderness. Twelfth-century millers usually had typically Norman names (such as William), even though many of the individuals were actually Saxon by birth. A similar upwardly mobile trend was found among physicians, but other tradesmen, such as carpenters, generally had Saxon names.

The controversy over the Drayton windmill arose when Walter Fitz Robert was personally supervising the Daventry lordship. While his mother lived, the area had been under her control, and her second husband, Saher de Quency I, had even styled himself the lord of Daventry. After Walter had managed the manor for a time, he gave it to one of his sons, as part of his policy of providing for all of them. The baron lived so long that he found it necessary to bestow holdings on his children during his lifetime, rather than have them wait until his death to inherit the lands. Yet he also had to preserve the family patrimony intact. Therefore, he decided that Daventry would formally pass to Robert, his eldest surviving son, but that Robert should immediately grant it to his half brother, Simon Fitz Walter (Matilda de Bohun's son), as a fief. These arrangements were made before 1196—perhaps long before. Walter further required that Robert and Simon each confirm Daventry Priory's possessions, including its three mills, in words almost identical to those of the charters executed by Walter and his mother, Matilda of Senliz. Simon did not hold the fief very long, however. Upon his untimely but undatable death, Daventry passed to his son, Walter Fitz Simon.[32] This Walter was to prove as enthusiastic a windmill champion as his grandfather.

In the thirteenth century Roger the miller supervised many Daventry milling operations, including some mills owned by Walter Fitz Simon. For example, Prior Nicholas (c. 1231–1264)

---

32. B.L. Cotton MS. Claudius D XII, fols. 7v–9.

16, 17. A post-mill has been converted
into a catapult, used to hurl beehives
at a besieged castle. These crude, unpainted
sketches were drawn about 1326.
From Christ Church College MS. 92
(*The Treatise of Walter de Milemete*),
fols. 74v, 75; reproduced by permission
of the Master and Fellows of
Christ Church College, Oxford University.

observed that Walter Fitz Simon had once given the monks a windmill in Daventry whose profits were two shillings each year. Roger the miller operated it.[33]

In 1244 Prior Nicholas and his monks decided to renegotiate their contract with Roger, who was managing several of their

33. Ibid., fol. 9 (*molendinum ad ventum*). Between 1289 and 1291, Walter Fitz Simon's son, Robert, indicated that his family had a windmill in the west field of Daventry; *Christ Church Cartulary*, p. 78. Whether this is the same mill Walter gave the monks, or yet another post-mill, is unclear.

mills, as well as those of Walter Fitz Simon. The monks released him from the duty of paying tithes on the profits of three mills in Daventry that the miller held from them—two water mills called Middle Mill and Cuttle Mill and an unnamed windmill. Henceforth, Roger should pay a rental of twelve pence. Whether this modest sum was the total rent, or (more probably) the individual rental, is not clear.[34] Roger quitclaimed to the monastery any rights he had with regard to two other mills—a horse mill (a type of treadmill turned by animal power) in the priory courtyard and another water mill by the lower stock pond. Furthermore, he agreed to pay the Daventry monks a tithe of six pence for the windmill in Drayton he leased from Walter Fitz Simon. The miller sealed the agreement with his own signet, a white disc with a fleur-de-lis and his name inscribed around the edge.

Roger was thus an extraordinarily successful miller, for he supervised the operation of at least three water mills, two windmills, and a horse mill. Fifty years earlier, three different millers had been active in the same area. He probably employed subordinates, for he could not operate all the machines simultaneously himself. This was a family enterprise; Roger subsequently entrusted at least two water mills and the Drayton windmill to his sons, first Geoffrey and then John.

John was a carpenter as well as a miller. During his long lifetime (he lived well into the fourteenth century) he lost control of one windmill. Between 1289 and 1318, he received a report affirming that his father, Roger the miller, had held from the monks an acre in the south field of Drayton on Ryenhull, as well as the windmill there, Ryen Hill Mill.[35] This report is further testimony to the durability of windmills.

---

34. *Christ Church Cartulary*, pp. 54–55.
35. Ibid., pp. 56–57. The report was issued during Prior Peter's tenure, 1289–1318. For more on John, see pp. 85, 119.

It is now evident that by the mid-thirteenth century, the milling registry of Daventry listed the following:

A horse mill in the priory courtyard
A priory water mill by the lower stock pond
A priory water mill called Cuttle Mill (this mill lasted until the early sixteenth century)
A priory water mill called Middle Mill
A priory windmill in Daventry
A priory windmill in Drayton called Ryen Hill Mill
A windmill of Walter Fitz Simon in Daventry, later given to the priory
A windmill of Walter Fitz Simon in Drayton

Of the seven mills that belonged to the Daventry monks, three mills in Daventry (including Middle Mill and Cuttle Mill) were part of Matilda of Senliz's initial grant. Since the horse mill stood in the monastic courtyard, it was undoubtedly not given by someone outside the priory. The water mill by the monks' stock pond was also probably not an external grant. That leaves only the priory windmill in Daventry as the third mill donated by Matilda. Until 1244 it was simply referred to in the records as a mill. The language Prior Nicholas then used to refer to Middle Mill, to Cuttle Mill, and to this unnamed windmill was strikingly similar to that employed repeatedly by Matilda, Walter, Robert, Simon, and Walter. All this strengthens the contention that Matilda's initial grant, made about 1158, included a post-mill and that her son, her grandsons, and her great-grandson all upheld her original donation.

The priory windmill in Drayton, called Ryen Hill Mill, must be the same machine that caused a controversy when it was illegally erected by William of Drayton, surrendered to Walter Fitz Robert, and given by him to the monks. That mill, as well as the two not owned by the monks (Walter Fitz Simon's windmill in

Daventry, first noted before 1231, and his post-mill in Drayton, first mentioned in 1244), were once operated by Roger the miller. Walter Fitz Simon's two wind machines cannot yet be verified as existing in the twelfth century, although it is likely that they did.

The history of this one family's windmill sponsorship is worth recapitulating. Matilda of Senliz gave Daventry a windmill after 1158. Her son, Walter Fitz Robert, confirmed that grant before 1166. When one of his vassals brashly erected a competing post-mill in Drayton, Walter confiscated it and gave it to Daventry priory. Walter also gave land for a windmill at Hempnall to the community of Bury Saint Edmunds. His second wife, Matilda de Bohun, had a windmill on lands in Henham that Walter gave her at their wedding. Her son by her first marriage built one at Cleydon. Walter's son, Walter, may have had something to do with a post-mill in Hertfordshire. Baron Walter's grandson, Walter Fitz Simon, gave a windmill in Daventry to the monks and erected another one in Drayton for himself.

This brief overview does not include any Clare water mills. Nor does it cover all the windmills built after 1200. The members of each generation liked to take full credit for a gift, even if they were only confirming an earlier donation. It is therefore conceivable that there is a slight overlap in the Daventry-Drayton assembly. Whatever the exact total (and it was probably larger than suggested here), it represents a remarkable achievement in terms of charity and technology for one family.

There is an interesting postscript with regard to the Clare patronage. Walter Fitz Robert gave Dunmow Priory most of the land and ecclesiastical rights in tiny Hempnall, but he gave the Twigrind water mill and the land in Longbridge on which to erect a windmill to Bury Saint Edmunds. The priory canons subsequently complained that this large abbey was continually depriving its Hempnall parish church of the tithes derived from the two mills. In a surprise concession, the monks of Bury, temporarily led by saintly Abbot Hugh (1213–1229), agreed that the Dunmow canons were indeed entitled to these tithes.[36]

The controversy over who should pay and who should receive windmill tithes would become almost as heated as the quarrel over the construction of one mill in proximity to another. However, before analyzing the additional problems of tithing (Chapter 7), the case for the antibaronial proponents of the unrestricted use of air power should be presented.

36. B.L. Harley MS. 662, fol. 96v.

# Free Benefit of the Wind

"*T*ear it down! Tear it down!" yelled the black-robed abbot. Eight hundred years later, the prelate's fury still ignites the most volatile account of a contested windmill franchise. The scene of the dispute was the powerful Suffolk monastery, Bury Saint Edmunds, but the issue concerned people everywhere.

That large religious house numbered among its monks a talented historian and graceful writer, Jocelin of Brakelond, who was blessed with keen powers of observation and a retentive memory. The principal focus of his writing was the imperious Abbot Samson, who ruled the Bury convent from 1182 to 1211. Brother Jocelin was a shrewd but ardent admirer of his father abbot. His vivid descriptions are extremely valuable because he was an eyewitness to the events he described and because he wrote his account while they were still fresh in his mind; he finished the extant portion of his work sometime before 1205. Although his one anecdote about a windmill constitutes only a small part of his chronicle, it does epitomize his purpose and style, and does highlight an important facet of early post-mill experience.

According to Jocelin, an elderly rural dean named Herbert

erected a post-mill on his own land at Haberdun. This is a section on the outskirts of Bury, near the town's south gate, about three-quarters of a mile from the abbey church. Upon learning about the mill, Abbot Samson flew into a terrible rage and would scarcely eat or speak. The next day, after celebrating mass, he abruptly ordered the abbey sacrist to have his carpenters tear down the offending mill without delay and place its timbers in safe-keeping.[1]

Having been informed of the abbot's reaction, probably by a friend in the religious community, the dean voluntarily came to Samson and forcefully protested the stern edict, arguing that the free benefit of the wind ought not be denied to any man. Somewhat more apologetically, Herbert declared that he did not want it thought that he was competing with other mills in the area; he only wished to grind his own grain, not that of someone else. Samson furiously retorted, "I thank you as I should thank you, if you had cut off both my feet! By God's face, I will never eat bread till that building be thrown down. You are an old man, and you ought to know that neither the king nor his justiciar can change or set up anything within the liberties of this town without the assent of the abbot and convent."

Outbursts of bellowing rage were quite fashionable in the late twelfth century and Samson was no mean practitioner. He, and everyone else, had an illustrious model. King Henry II was renowned for his monumental, carefully calculated fits, and some of his subjects undoubtedly patterned their explosions after his.

Abbot Samson was a faithful monk and a conscientious leader, but he was essentially a business-minded individual. He believed it was his task to recover those possessions and privileges of his house that had been squandered by previous monastic administrations. Samson was determined to protect the abbey's remaining traditional rights, and he regularly insisted that nothing was to

---

1. This incident is reported in Jocelin, *Chronicle*, pp. 59–60. For a physical description of the abbot, see pp. 39–43.

be done in Bury Saint Edmunds or its environs without his express permission. His demand was strengthened by historical precedent because, as he claimed, the abbey did enjoy unusual independence from royal officials within what it called its liberty—a large area of some eight and one-half hundreds comprising most of western Suffolk.

Samson had a genuine commercial objection to Herbert's mill. Thus he thundered at the dean: "Why have you presumed to do such a thing? Nor is this done without detriment to my mills, as you assert. For the burgesses will throng to your mill and grind their grain there to their heart's content, nor should I have the lawful right to punish them, since they are free men. I would not even allow the cellarer's mill, which was built of late, to stand had it not been built before I was abbot."

Jocelin (and, apparently, Samson) did not use very precise language when referring to additional mills. At first, Jocelin specifically identified the mill at Haberdun as a windmill (*molendinum ad ventum*), but later in the same passage he called it simply a mill (*molendinum*), the term he also used in referring to the cellarer's facility and others throughout the book. The author's use of these terms interchangeably in one passage raises the familiar possibility that several other abbey mills were wind-powered. Bury did possess four verifiable windmills, the one at Hempnall that Walter Fitz Robert had given the monks and others at Elveden, Risby, and Timworth.

Samson was primarily concerned about the mills in the immediate vicinity of the abbey, of which there were six. One of them, Spinthil Mill, was located at the top of a hill; therefore, it must have been a windmill. However, neither its chronology, nor the identity of the cellarer's new mill, is clear.[2]

In any case, the abbot roughly dismissed poor Herbert. "Go

2. For an analysis of Bury (including maps) at a later date, see Robert S. Gottfried, *Bury St. Edmunds and the Urban Crisis, 1290–1539*, pp. 18–19, 77–79; maps 2–4.

away! Go away!" he shouted. "Before you reach your house, you shall hear what shall be done with your mill." The chronicler would have his readers believe that the old dean was terrified and fled the scene; quoting the prophet Jeremiah, he described Herbert "shrinking in fear from before the face" of the abbot.

In fact, Herbert was remarkably resilient and he immediately sought the counsel of one of his sons, Master Stephen, a man of considerable importance in his own right. The two agreed that the abbot was not about to relent. Aware that the carpenters were fast approaching, they ordered their own servants to tear down the newly erected windmill. When the sacrist's workmen arrived, they found nothing left to demolish.

This is a good story, full of character insight and practical details about mill craft. To Jocelin, Samson was a figure of immense fascination, but he was not uncritical of him. The monastic historian was primarily interested in two things: the steps his clever abbot took to reorder the abbey's finances, and the divisions such necessary measures created within the Bury religious community. For the monks, the Haberdun incident was a minor event in their complex domestic campaign. For windmill enthusiasts, it raises a larger issue.

Abbot Samson was in his late forties when he called Dean Herbert an old man. He was a first-rate administrator, but he was also hard, tactless, and rather secretive. A Norfolk native, he always retained a pronounced regional accent. Born at Tottingham in 1135, he had a fine education, including university study in Paris. As Master Samson, he had taught school before entering Saint Edmund's monastery at age twenty-four. When he was elected abbot more than two decades later, in 1182, he was the subsacrist—not a very exalted position, but one that allowed him to gain considerable insight into the recent mismanagement of the abbey's finances.

Just before the election, there had been tension between those monks who wanted a scholar as their next abbot and those who wanted a more consciously spiritual leader. The victorious Samson certainly had ample scholarly training, if not inclination, but on the whole, he was a rather pragmatic, worldly individual. On one occasion he opened and reverently inspected the tomb of the martyred abbey patron, Saint Edmund (d. 870). Yet the treatise he wrote on the miracles of Saint Edmund was hardly the product of original thought, for he borrowed heavily from earlier writers on the same subject.

Jocelin described Abbot Samson as being of medium height and quite bald. He had a prominent nose, thick lips, penetrating gray eyes, and shaggy eyebrows that he shaved frequently—surely a harmless but eccentric vanity. He also had a long red beard that turned to snow during his abbacy, was very fond of sweets, and loved to ride. Samson was an energetic traveler and also something of an exhibitionist; he once arrived fully armed and mounted to lead his knights to join the royal host. He loved display and was responsible for the construction of many buildings, including a huge tower at the western front of the abbey church and a hospital at Babwell. He also encouraged continuation of the abbey's splendid literary and artistic traditions, inspired perhaps by the Bible luminously illustrated by Master Hugh about 1150 and now preserved at Corpus Christi College, Cambridge.

After the conquest of Jerusalem by the Moslems and the proclamation of the Third Crusade, Samson continuously wore a coarse shirt and haircloth drawers. He also abstained from eating meat. Samson had few close friends, but he sought the opinion of the house physician, Master Walter, on a variety of matters. Jocelin claimed that although the abbot had a great affection for certain individuals, he never allowed it to be revealed by his countenance. Of course, this was a prudent policy for the abbot of a large monastery. In sum, Samson exhibited in full measure the contradictory traits that were so typical of Anglo-Norman people. He was at once religious and worldly, placid and militaristic, generous and intolerant.

Like his predecessor abbots, Samson lived in a luxurious state in an establishment a bit apart from his fellow monks. He was indeed their stern, unsmiling spiritual father, but he was also the lord of an immense fief, the so-called liberty of Saint Edmund. He had some fifty-two military vassals who did him homage for enfeoffed lands, and he was personally responsible to the king for the services of forty armed and mounted knights. Besides these obligations to his sovereign, he had responsibilities to his many different dependents. He dispensed justice to his manorial tenants in a rough and ready manner, once likening his idea of fairness to the maxim that "he who comes first to the mill, ought to have first grind."[3]

Samson's main problems, however, were financial. His monastery, like many English religious houses at that time, was deeply in debt. His less tenacious predecessor, Abbot Hugh (ruled 1157–1180), had frequently allowed valuable lands and benefices to slip out of abbey control or to be held for inadequate returns. Moreover, the convent had regularly spent excessive sums for the lavish entertainment of important guests. This had encouraged a lowering of the monastery's standard of spirituality. The financial situation was complicated by the fact that the total endowments were legally divided between the abbot and the convent. One advantage of this arrangement was that during an abbatial vacancy, such as had preceded Samson's election, only part of the temporalities reverted to royal control. However, a disadvantage was that the monks could resist whatever they considered interference in their internal affairs, even if such action had been instigated by the abbot himself.

Samson refused to tolerate unpaid debts, exorbitant expenses, continued alienation of property, or excessive division of authority, and his high-handed attempts to reestablish the abbey's solvency offended many monks and townspeople. One of his less

3. Jocelin, *Chronicle*, p. 56.

invidious techniques was to catalog some of the monastery's revenues and rights and to consult the summary whenever possible.[4] A systematic survey of the abbey's possessions was never completed, however.

The monastery's creditors included both Jewish and Christian moneylenders. Although the Jews were legally guaranteed special immunities and protection by the crown, Hebrew communities throughout England encountered mounting hostility in the second half of the twelfth century. As borrowing by Christians increased, they began to hate their usurious bankers. Some even made wild accusations that Jews were annually torturing innocent Christian boys in a perverted commemoration of Christ's Passion.

Some Crusader zealots argued that it was absurd to fight an enemy half a world away in Palestine, when some of his Jewish allies might be exploiting the neighbors at home. In 1190 atrocities were committed in several towns, including York, the site of a massacre and a mass suicide of Jews. Abbot Samson took advantage of the widespread hysteria to expel many Jews from Bury Saint Edmunds. His reasoning was strikingly similar to the basis of his argument against Dean Herbert. He resented the Jews' unique independence of local authority and was resolved to establish his jurisdictional supremacy. The main causes of both the Haberdun confrontation and the Jewish persecution were thus money and power. Samson was only partially successful, however, for years later he was still in debt to Jewish bankers and a few Jews were still living in the town of Bury.[5]

There is a possibility that Samson's intolerance may have been preserved in an unusual way. According to the interpretation of one former museum administrator, the magnificently carved ivory

---

4. For an example of his cataloging, see Samson, *Kalendar*.

5. There are many important studies of the English Jewry. For overviews, see Joseph Jacobs, ed., *The Jews of Angevin England: Documents and Records from Latin and Hebrew Sources*; Henry G. Richardson, *The English Jewry under the Angevin Kings*; Cecil Roth, *A History of the Jews in England*.

crucifix that is now in The Cloisters in New York City may have originated in Bury Saint Edmunds. This intricate masterpiece was apparently created about 1150. Approximately three decades later, Abbot Samson supposedly caused a series of anti-Semitic verses to be inscribed on the cross and had an image of himself inserted in one of the upper panels. This sounds like something Samson might have done, but whether he in fact did must remain pure conjecture. Other interpretations suggest a Canterbury or a Winchester origin for the crucifix.[6]

At Bury Saint Edmunds, the abbot and monks had a tested organizational plan for achieving their fiscal objectives. Both the abbey sacrist and the cellarer played significant roles in the management of the convent's finances. The sacrist was in charge of the service of the altar, including the costly lighting of the church, but he was also responsible for general building construction and repair. Through his prefects he also ran the town of Bury; any revenue they collected was allocated to the maintenance of the monastery church. In Samson's time this income went into a fund for erection of the west tower.

The cellarer was the abbey's general business agent; he was in charge of purchasing food and supplies for its huge membership. He also administered lands in the vicinity of Bury and used their profits for the normal upkeep and domestic operation of the monastery. Clear as this division seems, manifold opportunities existed for conflict between the two officials, particularly over the revenues from the *bannaleuca*, or belt of land immediately beyond the town walls.

Part of this disputed area was called Haberdun and consisted of the acreage between Southgate Street and the River Lark. Although this land was strictly within the sacrist's domain, as Samson's demolition order with regard to Herbert's post-mill indicates, it appears that the cellarer had successfully invaded other

6. Thomas Hoving, *King of the Confessors*, esp. p. 331.

nearby spots and had recently built there. Dean Herbert, a man knowledgeable about local affairs, may have hoped to capitalize on these petty jurisdictional rivalries when planning the construction of his windmill.

The dean was a stubborn man of spirit and substance who was not easily defeated. His father, Robert, was related to Abbot Ording (1138, 1148–1156), and that kindly prelate had been instrumental in furthering Herbert's career. Herbert was first given a chaplaincy at Rougham, next to Haberdun, presumably by the bishop of Norwich. Abbot Ording offered him some land (which eventually totaled seventeen acres) in Rougham adjoining the house of Herbert's uncle, Solomon, who was also a cleric. Gradually he accumulated other small plots, including land he held in fee from the same uncle.[7] Herbert became dean during the tenure of Abbot Hugh (1157–1180), whose writs he had occasionally attested.[8]

Even though he was born well after the church had launched its reforming campaign against clerical marriage, Herbert, like many country priests, managed to raise a family. He had at least two sons. Adam received some family property within his father's lifetime. Master Stephen, a clerk and university graduate, became an active member of Abbot Samson's household and eventually held a valuable benefice at Barton Mills.[9]

Herbert's ecclesiastical title has been variously interpreted (some say it was purely honorific). He was evidently a rural dean, that is, a diocesan official sometimes known as an archpriest.

7. On Herbert's lands, see Douglas, *Feudal Documents*, pp. 131–32, charters 140–41 (1148–1156); Samson, *Kalendar*, pp. xxxix, 18–20.

8. For his attestations, see Douglas, *Feudal Documents*, p. 137, charter 147 (1156–1180); p. 140, charter 151 (1164–1180); p. 141, charter 152 (1156–1180); p. 149, charter 166 (1166–1180). See also Samson, *Kalendar*, pp. 164–65, charter 154 (1182–1211).

9. Stephen's name appears in Samson, *Kalendar*, charters 1, 14, 18, 26, 34, 36, 37, 56, 57, 59, 60, 85, 93, 106, 107, 111, 119. On his land at Barton Mills, see charters 36–37.

Rural deans were ultimately responsible to the bishop, in Herbert's case the bishop of Norwich, not to the abbot. They administered subdivisions of an archdeaconry, often a shire hundred or a group of hundreds. Although they followed archdeacons in ceremonial precedence, rural deans had considerable freedom of action. Their duties included directing the parish clergy within their deaneries, supervising elementary education, conducting trials by ordeal, denouncing and impleading notorious sinners, distributing chrism and other holy oils, and enforcing papal decrees against unchaste priests.[10]

This last task must have occasioned irony and derision when Herbert assumed it, for assigning him to punish married priests was like asking the wolf to guard the chickens. Because deans, like archdeacons, frequently dealt with matters of law, finance, and sexual morality, they faced any number of corrupting temptations. King Henry II tried to eliminate abuse of their power by forbidding them to initiate unsubstantiated prosecutions and by directing that they first take complaints to juries of presentment.

In the Norwich diocese, a dean's problems, or temptations, were far less obvious because tradition had made the rural deanery into a freehold benefice, that is, a sinecure, which required little active service. When Herbert claimed that Haberdun was a free fief (a seemingly contradictory term because fiefs were normally held by contractual agreement in return for some sort of service), he may have been referring to the fact that he held it as an ecclesiastical benefice from the bishop, not as a feudal tenement from the abbot. The records of Bury Saint Edmunds clearly show that the dean owed payments to the abbey for other area holdings. Thus like many men, he was serving two masters.

Such distinctions did not impress the abbot when his own prerogatives were being threatened; he did not even respond to

10. Rural deans are discussed in Frank Barlow, *The English Church*, 1066–1154, pp. 80, 137, 155, 159, 227; Alexander H. Thompson, *The English Clergy and Their Organization in the Later Middle Ages*, pp. 49–50, 63–67.

the dean's definition of his holding. If Herbert's claim that he had the right to build a windmill on his own free fief were upheld, the consequences could have been enormous. But Samson asserted that he had jurisdiction over everything that happened in the liberty of Bury Saint Edmunds. And he prevailed.

Herbert's learned son, Master Stephen, could offer no legal solution, but he did have some personal connections, including a number of influential people, such as the abbey steward, Robert of Flameville, with whom he frequently associated in various types of transactions. One such association involved the Arsics and the Bricetts. The dean knew Reginald Arsic's lord, Aubrey de Vere, and also the prior of Hatfield Regis.[11] Whether these friendships had sparked Herbert's interest in windmills cannot be determined, but the case of the outspoken dean is a reminder that technological progress and a desire to initiate commercial enterprise were not confined to the young.

Many scholars have confidently asserted that Herbert's quarrel with Abbot Samson occurred in 1191 or 1192. Their reasoning is based on the fact that the preceding event in Jocelin of Brakelond's chronicle can be dated 1191 and the succeeding event, 1192. Although it seems logical to assume that the windmill controversy occurred between these two episodes, such a conclusion is not necessarily correct. Jocelin's narrative does generally, but not invariably, follow a chronological sequence. He is obviously incorrect about the chronology of some of the incidents he describes. Moreover, there is no indication in his account that the Haberdun confrontation definitely took place between 1191 and 1192.

In fact, just the opposite seems true. Samson complained that the cellarer had improperly built a mill "of late," under an earlier administration. That would mean before 1180. It hardly seems

11. Cambridge University, King's College MSS., Bricett Deed B 10; *Colchester Cartulary*, 1:150–53.

likely that he would have used the phrase "of late" a full dozen years after the event. Moreover, the tone of the whole encounter resembles that of the first years of Samson's abbacy, perhaps between 1182 and 1184, when people were still testing his resolve. Thus, a date of 1191 or 1192 for the construction of Dean Herbert's windmill seems altogether improbable.

The quarrel between Herbert and Samson had no major repercussions in the Bury area. To the abbot's credit, he bore no grudge against Herbert and continued to employ Master Stephen, even advancing him to a higher position. Windmills multiplied, but in areas slightly further from the town and only under the abbot's license. Samson's objection to new mills that were erected to the detriment of existing ones is strikingly reminiscent of Walter Fitz Robert's strictures against mill competition in Daventry, Drayton, and Hempnall.

Abbot Samson definitely knew of other windmills besides those at Haberdun and Hempnall. The contract by which he leased to Solomon of Whepstead the fully stocked Breckland manors of Elveden and Ingham included a detailed inventory of the equipment, livestock, and produce of the two estates. It indicated that Ingham had two profitable water mills with all their apparatus (*duo molendina ad aquam pacabilia cum tota apparatu suo*), and that windswept Elveden had a good post-mill, also with its apparatus (*molendinum unum bonum ad ventum cum apparatu suo*).[12] The lease, whose term was Solomon's lifetime, stipulated that for Ingham he would pay a monetary rent and that as payment for Elveden, he would deliver a specified amount of produce each year. The latter arrangement ultimately proved to be a wiser policy since it took into account fluctuations in market prices.

The abbot's terminology was undoubtedly quite clear at the time, but now it lends itself to several interpretations. Some of the language in this lease was probably the rhetorical flourish of

12. See Appendix entry no. 43.

an abbey scribe. In other charters from Bury Saint Edmunds, mills were given nicknames, or described in terms of their location. In this instance Samson, or his copyist, was probably not deliberately contrasting two profitable (or serviceable, or ratable) water mills with a good windmill, but rather simply specifying that all mills functioned correctly.

Nevertheless, the word "good" might have additional connotations, such as in working order (possibly implying that other, unmentioned mills at Elveden were not), or profitable, or improved, or even legal (an interesting possibility in view of the dispute with Dean Herbert). The description "with its apparatus" indicates that the lease covered more than just the mill building and that the installation already existed—it was not merely in the planning stage as were the windmills described in some documents.

Who was the new leaseholder? Solomon was a somewhat unusual name in Angevin England. Although there were numerous Jews in the town of Bury, the name was not confined to men of that faith. Old Herbert's Uncle Solomon, for example, was a cleric. That uncle might have been the same man as Solomon of Whepstead. The dean's advanced years make such a coincidence unlikely, but not impossible. There may have been yet another similarly christened burgess in Bury, Solomon of Muckings. However, this individual also held property at Whepstead and thus can probably be identified as Solomon of Whepstead.[13]

The date of Solomon of Whepstead's lease pertaining to Ingham and Elveden can be placed only between 1182 and 1200. The decision would seem to have been made in the first years of Samson's abbacy. Between about 1180 and 1220, the price of English grain and a few other commodities skyrocketed. This unexpected rise followed decades of price stability in Western

13. On Solomon of Whepstead, see Samson, *Kalendar*, charters 23, 26, 58, 60, 106; *Feet of Fines, Suffolk*, items 318, 371. On Solomon of Muckings, see Samson, *Kalendar*, charters 62, 63.

Europe.[14] Some modern economists have compared this inflation with similar economic phenomena that occurred in the sixteenth and twentieth centuries—also eras of outstanding commercial and industrial achievement.

As in any inflation, individuals and institutions dependent upon fixed incomes suffered the most. Landowners discovered that long-term tenants, who paid a set rental, were suddenly reaping greater profits than they were. Thrifty lords, such as Abbot Samson, increasingly opted for direct management of their estates. Samson did not always wait for default, expiration, or repurchase of old leases. In his first year in office, he simply canceled or renegotiated a number of old covenants.

The preferred policy was then to refurbish the manor and either administer it directly as a grange under a few monks who were sent to live there, or re-lease it for a shorter period with a higher rental that usually took the form of payment in kind. Since Solomon of Whepstead was offered a lifetime lease (with a flexible rate of payment for Elveden), it is likely that the terms of his contract were set early in Solomon's tenure, when the abbot was insisting on manorial renewal, but before he had adopted the strategy of short-term leases.

The abbot's policies were seldom consistent, however. About 1160, Elveden and Ingham had been leased for only a twelve-year period;[15] yet strangely, Samson later drew up a lifetime lease for Solomon. Well into the 1190s he was still allowing other lifelong covenants. If Solomon of Whepstead was an elderly man, as Dean Herbert's Uncle Solomon surely must have been, then a single life-term contract could have been, in effect, a courteous short-term lease. It would also have been a wise maneuver, if the intention were to prevent the estate from passing to Solomon's heirs, even temporarily.

14. P. D. A. Harvey, "The English Inflation of 1180–1220."
15. Douglas, *Feudal Documents*, p. 135, charter 145. Ralf Brian rented the manors.

The name Solomon has another fascinating link with wind-mills. The familiarity of Bury Saint Edmunds with mechanized power may have inspired one of the earliest known paintings of a post-mill. In the thirteenth century (probably about 1270), a talented East Anglian illuminator, perhaps a monk from the Bury community or surrounding area, included such a windmill in an unusual illustration that, spread over two facing pages, introduces a collection of the psalms of King David. This is one of the finest representations of a medieval post-mill; it is not a casual or prim-itive sketch, but rather a complex rendering by an accomplished artist. The depiction of the milling facility has caused the whole manuscript to be popularly referred to as "The Windmill Psalter" (see Figure 18).[16]

The windmill scene is really an intricate elaboration of the letter *E* of the word BEATUS, with which the first psalm com-mences. That poem begins, "Blessed is the man who hath not walked in the counsel of the ungodly." It continues:

> And he shall be like a tree which is planted near
> the running waters, which shall bring forth its fruit, in
> due season.
> And his leaf shall not fall off: and all whatsoever
> he shall do shall prosper.
> Not so the wicked, not so: but like the dust,
> which the wind driveth from the face of the earth.[17]

The whole psalm sings of wisdom and of justice, and its East Anglican expositor tried to make the text more vivid by using specific, if allegorical, images to illustrate it.

On the left-hand page, within a large letter *B*, Christ in Maj-esty is depicted springing from the Rood of Jesse. This was a

---

16. Pierpont Morgan Library, MS. 102, fols. 1–2. For a full-color reproduction of the windmill, see Jonathan A. Alexander, *The Decorated Letter*, plate 30.

17. *The Holy Bible*, Douay-Rheims Version, New York: P. J. Kenedy and Sons, 1950, p. 581 (Psalm 1, vv. 3–4).

18. The letter *E* in *BEATUS* frames the judgment of Solomon.
The post-mill and curling leaves may suggest blowing wind,
indicating the presence of God. Painted by an English artist
between 1270 and 1290. From MS. 102 (The Windmill Psalter),
fol. 2; reproduced by permission of the Pierpont Morgan Library.

traditional iconographic symbol, used to represent the ancestry of
the Redeemer. A large letter *E* appears in the top half of the right-
hand page. This format is unique because usually only the first
letter of a word was historiated. In this illustration the central
arm of the *E* supports David's son, King Solomon, who is seated

in judgment, about to decide which of two grieving women is the real mother of an infant. He holds upright the great sword of justice. A soldier stands ready to execute Solomon's dreadful command, which is to split the child in two.

Beneath the Israelite king, in a dazzling display of technical skill, an angel is depicted flying downward carrying a scroll bearing the initial word's remaining letters, *A-T-U-S*. This twisting descent seems to prefigure similar flights from later Venetian frescoed ceilings. The angel's wings are spread, his cloak is flowing, and he seems to stir the air by his passage; vine leaves throughout the panel curl as though caressed by a strong breeze.

Beneath the heavenly messenger, the text of the psalm continues, punctuated by a fanciful representation of a bald man with an animal body. In the bottom quarter of the page stands a splendid solitary peacock.

Far above the bird, the animal man, the hurtling angel, and the pensive Hebrew king, in the swaying green foliage atop the letter *E*, serenely sits a post-mill. It is about two inches high and is viewed from the back; the four sails are shown on its far side. The cabin, or buck, rests on tripod supports and a ladder leads down from the entrance. Visually, the windmill provides a crowning fifth level of activity in the painting; ideologically, it unites the artist's entire conception with the Hebrew poet's prayer.

Mills grind grain into valuable flour and scatter waste particles as dust. The post-mill may therefore allegorically exemplify the contrasting fate of the two mothers, the wicked and the just. Some exegetic commentators have asserted that the true mother in Solomon's judgment represents the church and that the pretender represents the synagogue, thus interpreting the king as a prefigurement of Christ the Judge. Other preachers and scriptural scholars have viewed the episode as exemplifying the Last Judgment, which will eternally divide the elect from the damned.

Unfortunately, neither the artist nor the donor of this masterpiece is known. Certainly, the latter was a person, or foundation, of wealth and consequence, for the complete psalter must have been an extremely expensive acquisition. It is a large volume intended for public display or communal liturgy rather than for private devotion. One modern analyst, intrigued by the prominence give the letter *E*, has perceptively suggested that the psalter might be connected with someone at the court of Edward I (1272–1307), a monarch who prided himself on his attention to judicial affairs and who was sometimes called a second Solomon. This conjecture is consistent with the estimate that based on the style the manuscript should be dated to the 1270s.[18]

It is tempting to think that the twelfth-century Solomon of Whepstead who leased the Elveden windmill had a hand in commissioning this great work. He was a jurist of some local distinction. A representation of the judgment of Solomon, including the letter *E* (for Elveden) and a post-mill, would constitute a recognition of his technological, judicial, manorial, and spiritual interests. Evidently too many decades intervene between the Bury wind-power advocate and the painter of miniatures. However, it is entirely possible, even quite likely, that the artist who illustrated the Windmill Psalter derived his ideas from an earlier East Anglian manuscript whose purpose had been to praise the accomplishments of a man from Bury named Solomon, or of his son, who was a prominent windmill enthusiast. Graffiti depicting post-mills were scratched on church walls at Dalham and Lydgate, both of which are near Bury.[19]

Although the artist who illustrated the psalter exhibits a sophisticated style, he omitted the large tail pole by which the windmill could be turned around. This device was commonly neglected in graffiti sketches as well. In addition, the illuminator included three diagonal base supports but apparently forgot the main post. Of course, one must allow any artist considerable

18. Adelaide Bennett, "The Windmill Psalter: The Historiated Letter E of Psalm One."

19. Pritchard, *English Medieval Graffiti*, pp. 136–37, 149.

license. This one seems not to have been too interested in the building's engineering design. Strangely, some excavations do reveal a sunken tripod arrangement, without a central post.[20]

The very features that the psalter illustrator missed were the ones which most fascinated the English illuminators who decorated copies of Aristotle's *Meteorologica*. Although they were less accurate and skillful, their historiated initials were triumphs of action and emotion. These artists placed large, almost cylindrical bucks on stout central posts and deliberately exaggerated the tail poles. They also portrayed boys grasping the poles, apparently ready to push them around at any moment. One of the artists showed the mill sheets full face forward, omitted the bottom bracing supports (which were presumably buried) and left out the ladder (see Frontispiece and Figure 5).

Not until the fourteenth century did such painters as the superb miniaturist who illustrated the Luttrell Psalter depict all the windmill parts accurately (see Figure 14). However, another fourteenth-century illustrator, who included no fewer than six windmills in his edition of the *Decretals of Pope Gregory IX*, inserted an odd, cradle-shaped appendage underneath three of his post-mill cabins. The function of this structure has not yet been determined. One of his other post-mills almost seemed futuristic, except for a cock atop its upright sail (see Figures 19, 23, 24).

Countless artists in all countries, especially the admirable Dutch landscape painters, later found inspiration in post-mills and tower mills. Photographers were also captivated by their form and grace. The wonder is that not until more than a century after 1137 did artists realize that this subject had real potential. Once portrayed, however, it became a consistent romantic and whimsical favorite.

Similarly, the tale of Dean Herbert's windmill experience was slow to gain wide circulation. For more than six hundred years,

20. Jarvis, *Stability in Windmills*, p. 12.

only the monks at Bury Saint Edmunds, and later the local archivists, could read Jocelin of Brakelond's account. In the early nineteenth century, following publication of the Latin text and a colorful English translation of Jocelin's complete works, the noted historian Thomas Carlyle popularized many of the chronicler's best stories and devoted one long paragraph to the Haberdun incident. Typically, he interpreted the episode as another instance of the exercise of autocratic power by a great abbot, with old Herbert "tottering" away from a harsh ecclesiastical judgment. Always the stylist, Carlyle composed a good line about how easy it was "to bully down old rural deans, and blow their windmills away."[21] Today, his emphasis seems sorely misplaced.

Herbert had many characteristics of a true pioneer; he was a promoter not only of technical skill but also of free enterprise. In this, he resembled that other independent upstart, William of Drayton, who built the Ryen Hill windmill in Northamptonshire, only to have it taken away by Walter Fitz Robert. However, Dean Herbert framed his protest in larger terms. He lived life vigorously, taking risks and accepting consequences. In a highly structured society, he claimed he could do as he wished on his own free fief. In a tight little town, he attempted to grind his own grain without help from, and without owing payment to, anyone else.

Dean Herbert contended that the power of the free air should be available to everyone—a seemingly obvious but nonetheless revolutionary idea. The speed with which he erected and dismantled his post-mill proved that average people could harness the wind. The new technology therefore created a genuine "poor man's power"—a new way to use a remarkably cheap, unfettered resource. Its development and application represented a frontal attack on traditional baronial privilege.

Abbot Samson considered Herbert's action and his arguments

21. Thomas Carlyle, *Past and Present*, pp. 118–19.

dangerous precedents that could encourage other burgesses to attempt to circumvent the well-nigh universal requirement that flour be ground in a central, taxable place. In characteristic aristocratic fashion he therefore resolved to prohibit construction of any new mills except his own. His reaction was quite logical, in view of his definition of the proper social order as hierarchical.

On the manor, a lord could force his serfs to use his own mill. Sometimes even hand mills (querns), which ground small amounts of grain for personal consumption, were forbidden. However, in prosperous, heavily populated East Anglia there was a greater proportion of free tenants and small landholders than elsewhere, and they jealously guarded their privileges. Abbot Samson therefore explicitly confirmed the ancient right of peasants within the liberty of Saint Edmunds to use querns. This was a mild relaxation of manorial strictures. A woman probably spent about two hours of hard monotonous labor each day grinding enough grain to meet the needs of her family. Yet in the aggregate, such individual milling could result in a significant diminution of abbey revenues. The loss would be aggravated, if a readily available machine were to expand individual milling capacity.

This local milling right probably influenced Herbert's thinking, but his comments on it were not recorded. He did specifically, and perhaps disingenuously, renounce any commercial intentions. In fact, he had a more important point. He did not claim that he had found a better technique for doing what he was already legally entitled to do. Rather, he argued that he had a right to build as he wished on his own free fief.

It is interesting how often the word "free" was used during this confrontation. Abbot Samson ignored Herbert's eloquent defense of the free use of the air, just as he ignored his definition of a free fief. Samson's reaction was almost predictable. For him, the significant issue was not the grain, or the innovative technology, or the fief's privileges, or the free air, but rather that the dean had challenged the status quo. Old Herbert had neglected to seek his permission before embarking on a new venture. Furthermore, the abbot feared that a wind machine would encourage Bury towns-people to avoid the abbey's mills, which would result in a reduction of its revenues. Accordingly, he inflexibly reiterated that all salable grain was to be ground in the monastic mills.

The real aggressor in this brief struggle thus seems to have been not the imperious, business-minded, rather vain abbot, but the bold (and perhaps impish) dean. He challenged the abbey's traditional authority by building a windmill almost in the monastery's backyard. The fact that he was forced to tear it down is less important than the individualistic spirit he displayed.

Perhaps Herbert did stand as an incandescent champion of the innate rights of every individual. Or maybe he was just a lifelong malcontent who saw another small opportunity to oppose the existing order, as he had done when he had married illicitly. Perhaps, as he implied, he just wanted to do something for himself. Whether Herbert was motivated by a larger purpose or only by frustrated self-interest, the issues that concerned him were power and personal integrity. The windmill was only the manifestation, that is, the symbol, of his protest.

Whether he realized it or not, Herbert had fired an opening salvo in a long campaign—a veritable windmill war that would be fought and refought throughout Western Europe. Each time an entrepreneur dared to erect his own mill in violation of an established franchise, the echo of the dean's battle cry could be heard: "The free benefit of the wind ought not be denied to any man."

# Crosswinds in Kent

$W$ind is a capricious force. It can blow steadily forward, or it can whirl about, turning back upon itself. Windmill proprietors could sometimes be as erratic as the restless force they strained to harness. A few even worked at cross-purposes to their own best interests.

Some patrons were imaginative and altruistic. Many were as concerned about health care as they were about technological progress. Others were shrewd and self-interested. A few were devious and cowardly. Nowhere were these contradictory traits exhibited more clearly than Kent, the ancient spiritual heart of the kingdom, where pilgrimages to Canterbury attracted people of every place and station—rascals and rogues as well as perfect gentle knights.

The early Kentish wind-power enthusiasts mounted a colorful Chaucerian troupe of their own. Performing together, although not always aware of it, were the mighty archbishops, a corrupt soldier, a talkative but self-effacing prioress, a craven lawyer, an energetic abbot, town officials, estate managers, and a few obscure shareholders. There was even a contingent of proud millers. None of them told a side-splitting tale, however. The only types missing were a jolly innkeeper, a laughing lusty wife, and a good-hearted poet.

The few records that exist reveal a veritable growth spurt of windmills in Kent from about 1180 to 1220. As sleeping orchards throughout the county burst into bloom with the coming of spring, so did post-mills blossom there, seemingly all at once—in Canterbury, at Chislet, probably near Maidstone, and certainly along the Channel coast at Reculver, Thanet, and Romney. In this hallowed little corner of marshlands, hopvines, and salt pans, the seedlings of windmill technology and health care unmistakably bore fruit. Reaping energy from the air had become more than fashionable; wind power was even subsidizing nonmilling activities.

The Anglo-Normans were an exceptionally litigious people. Nowhere was this more evident than in Kent. Dean Herbert's suit had been swiftly, and rather arbitrarily, adjudicated in his lord's manorial court. Some decisions were appealed to the king's bench. If a churchman or a significant new issue were involved, the case would likely be taken as far as the papal curia. This happened with a few unusual windmill disputes. As it became increasingly clear that the potential revenue from the new apparatus might be enormous, both civil and ecclesiastical tribunals accepted more post-mill contests.

A relatively minor squabble on the Isle of Thanet, off the shire's northern coast, entailed some basic principles like those decided at Haberdun, but it was eventually settled in the royal court. The twists and turns in this case show how crucial legal delays could be and how highly some post-mill promoters valued their machines. They also reveal some of the differences between baronial and royal justice. Most of all, they reflect the tremendous growth of the bureaucracy that took place in the twelfth century. This case is unusually well documented and can be traced for at least five years (1199–1204) through the enrollments of the Exchequer, the decisions of the royal courts, and the original charters and cartularies of the Canterbury monks. It thus has considerable legal, social, and technological significance.[1]

The protagonist in this episode was William of Wade (or Wode, or Wood), sometimes also called William of Acol, the name of his township in Thanet. A knight of minor rank, he was ambitious, imaginative, quick, and persistent. William was a generally respected individual who was occasionally asked to testify at local jury inquiries. Yet he pursued his objectives without regard for others and was known to have broken contracts and to have deceived royal authorities. A few years after Dean Herbert's attempt to assert his independence, an attempt by William to build a windmill on his own free tenement at Monkton, which adjoined Acol, brought him into conflict with both church and state.

The monks of Christ Church cathedral priory had a valuable manor at Monkton and claimed rights over most of the area, but it is not clear who controlled the land in question. William may have built his mill on monastic land, as his opponents contended. Or he may have owned a free holding there, as he declared. Or he may have held land in tenure from the monks. Whatever the truth of the situation, the Canterbury brethren vigorously objected to William's post-mill, probably because they felt it threatened the local monopoly enjoyed by their ancient water mill.

When it became evident to them that the knight was not about to comply with their demand that he demolish his mill, they haled William into court. The monks' preliminary legal tactics are unknown, but it was in the royal court, rather than their manorial court, that a decision was rendered.

According to testimony the monks delivered as late as 1204,

---

1. The evidence concerning the Monkton controversy is discussed in Appendix entry no. 10.

they had taken William to court about the mill during the reign of King Richard, that is, sometime between 1189 and 1199. The parties had then reached a formal agreement. According to the monks, William Wade had put himself at their mercy and had promised to destroy his windmill by the third day after the following Easter. He had further agreed not to erect another mill without the permission of the prior and the monks. Such settlements usually entailed some concessions on each side, but the Christ Church plaintiffs did not admit having made any.

The fact that this quarrel reached the royal court and was not adjudicated in the local manorial court, as Dean Herbert's protest had been, was a measure of the priory's weakness, not of its strength. In normal times Christ Church could easily have punished any violator of its rights without royal assistance, but the last third of the century was a difficult period for the famous monastery. The cathedral priory had lost influence and suffered financial exploitation during the exile of Thomas Becket. Despite the wealth that poured into its coffers after the archbishop's martyrdom, its leadership was notoriously unstable. Eight different priors were elected in a sixteen-year period; two of them were subsequently expelled. The cathedral burned down in 1174, and its reconstruction absorbed the monks' energies for many years thereafter. Moreover, a protracted bitter quarrel arose between the community and two successive archbishops who wished to establish nearby a collegiate church of secular canons to assume some of the priory's archdiocesan responsibilities. In 1188 this dispute escalated into a series of communal riots. The number of resident monks plummeted from 140 in about 1190 to only 77 in 1207.[2]

Prior Geoffrey was elected in 1191, and although his tenure was also troubled, he remained in office for twenty-two years. It

2. R. A. L. Smith, *Canterbury Cathedral Priory: A Study in Monastic Administration*, p. 3.

was probably during one of the preceding intervals of weak leadership that William Wade raised his post-mill at Monkton. Certainly it existed by 1198.

William's dispute with the monks was influenced by the memory of an earlier encounter. During the previous generation, the priory had aggressively increased its property holdings throughout the county. Between 1153 and 1167 Robert Wade, presumably William's uncle, offered to sell to Christ Church about fourteen acres in Thanet, near the house of one Godwin the cornmonger. Having received a seemingly fair price (six marks and three shillings), he arranged for his brother, his brother's wife, and their children to agree to the sale at a public meeting in Monkton.³ Evidently they granted their consent more grudgingly than anyone suspected at the time.

The reference to Godwin is important, for it may be the first time anyone in England had been recorded as being a grain merchant. It also suggests that agriculture was thriving in Monkton and that small farmers and large manorial corporations were producing grain for trade, as well as for private consumption.⁴

Prior Wibert of Canterbury leased Robert Wade's land to Robert of Swanton, who contracted to pay a yearly rent of eight shillings and four pence—roughly 10 percent of the sale price. The prior clearly showed some favoritism in the transaction, however, for Robert of Swanton was the brother of Ernulf, one of the Christ Church monks. William Wade never reconciled himself to the loss of the Monkton acres. Consequently, in addition to raising his controversial windmill, he reoccupied the Wade ancestral land when the opportunity presented itself, perhaps at the death of Robert of Swanton. For some reason, the monks seemed less

3. William J. Urry, *Canterbury under the Angevin Kings*, p. 225 (Rental A, 1153–1167).

4. Norman Gras, *The Evolution of the English Corn Market from the Twelfth to the Eighteenth Centuries*, p. 163, thought the term "cornmonger" first appeared in 1204.

upset by this second affront than they were by the post-mill construction.

Prosecuting William in the royal court probably seemed to the monks like a welcome alternative to their own weakness. Yet it entangled both parties to the relatively minor Thanet dispute in the assertive morass of late Anglo-Norman jurisprudence. Predictably, something went wrong and the royal verdict was not enforced. One way or another, William Wade managed to stall and thus avoided dismantling his windmill. When King Richard was assassinated in April 1199, William realized that he might have a second chance. He crossed the Channel and sought out King John, then in revel at Poitou. Making no mention of his compact with the monks, he offered the king ten marks to confirm his right to build a windmill in Monkton. Royal permission was granted on December 31, 1199, and the payment of ten marks was duly recorded on the pipe roll at Michaelmas 1200. Thus, in what was clearly an exercise of opportunism and stealth, William overturned the earlier settlement. Perhaps the knight was already acquainted with John; he certainly seems to have taken his measure.

Both William Wade and the new king had seriously underestimated the resolve of the Canterbury monks. These religious were never ones to abandon a legal fight, even against a monarch as unscrupulous as John. If nothing else, time was on their side, for group resentment always seems more enduring than individual hurt. Moreover, the members of this religious community had had decades of practice defending their myriad privileges against infringements by laymen. They immediately brought another action against William. The same pipe roll of 1200 that recorded the knight's payment of ten marks also contains an entry indicating that the prior and monks paid the king ten marks of their own to have the court reopen their case against William and his windmill.

19. A fanciful fourteenth-century depiction of a giant
attacking a castle. The post-mill has a somewhat futuristic
appearance, partly because it was left unpainted.
From Royal MS. 10 E IV (Smithfield Priory Decretals),
fol. 89; reproduced by permission of the British Library.
Figures 23 and 24 show other windmills from the same manuscript.

William's reaction was to stall again. This time, utilizing the
formal legal weapons at his command, he pleaded that illness had
confined him to bed and asked to be excused. If the dates are
accurate, he obtained three court postponements: two at Michaelmas 1199 and another early in 1200. Proctors presented his ex-
cuses and several legal advocates acted on his behalf during numerous proceedings. The cathedral priory also hired spokesmen,
but Brother Elias was delegated special responsibility to shepherd
his community's case. The court usually allowed only three postponements, and William took advantage of them all.

In the meantime the monks had been sharpening their argument. They had begun to charge not only that William had violated their rights and had maintained the windmill in defiance of

a verdict from King Richard's court, but also that he had deceived King John by remaining silent about the earlier judgment. This was a serious matter and William Wade found it to his advantage to seek a private peace with the monks. In 1202/3 he appeared in the priory's court in Canterbury and, in his words, put himself at the priory's mercy in an attempt to settle their differences.

William prepared a statement in which he agreed to (1) surrender title to twelve and one-half acres formerly held by Robert of Swanton; (2) promise that neither he nor his heirs would ever build another windmill in Monkton, or anywhere in Thanet, without the express permission and goodwill of the monks. Furthermore, William promised that neither he nor his heirs would ever again use the charter he had obtained from King John at Nottingham, in any place or in any discussion, against the prior or the convent. Presumably, this royal decree was in addition to, or in consequence of, the one that was issued in France. William had his Canterbury declaration witnessed by seventeen men and appended to it his own wax seal, a rather commonplace round design consisting of four crossed feathers with his name encircling the rim.

Note what William Wade did not do. Stubborn to the end, he did not agree to tear down the Monkton post-mill. He did not admit that he had obtained King John's approval illegally. And he did not volunteer to destroy the 1199 royal charter. In essence, he proposed a compromise. The monks clearly thought the encounter significant, for they carefully preserved William's 1202/3 proclamation in its handwritten form and also had its text copied and prominently displayed in their cartularies. However, they did not forgo their right to appeal to the king's bench. In October 1203 Brother Elias obtained a firm date for a trial, to be held the next year.

The reopened case was heard in the Hilary term of 1204, but only after the monks had paid the impressive sum of twenty-five marks for a speedy judgment. Whether this included, or was in addition to, the ten marks previously tendered by them is impossible to determine. The government obviously profited from the litigation, for it collected a total of thirty-five to forty-five marks. King John seemed already to be manifesting the avarice that would contribute to his downfall.

After initial delays, King John's court conducted its proceedings expeditiously, probably because the monarch's honor was involved, as was a formal decree of his predecessor's court. After ordering that the records of the previous trial should be reviewed, John rendered judgment in the monks' favor. Specifically, he decreed that the parts of his own earlier (1199) charter that pertained to the windmill should be annulled, that the mill should be torn down, and that another should not be reerected without the approval of the prior and the monks.

Although John's motivation in reaching such a decision is unclear, one senses that he was not concerned about any issue relating to technology or commercial monopoly. He seems to have been angry that William Wade attempted to deceive him and that the monks exposed the plot. The priory never made clear why it so adamantly objected to the Monkton windmill. Perhaps it did not really object but was merely annoyed that William had not obtained its consent. Why such consent should have been required was never clarified; both the king and the knight declared that the mill was on William's tenement.

There is no proof that the windmill was ever destroyed. Because both sides had invested so much money in the litigation, it seems likely that the victorious monks either took possession of the mill, or negotiated an agreement with William whereby Christ Church held title to the mill, but rented it back to him, or took a share of his profits. This was a common result of suits of this type. A contemporary rental agreement between the monks and Gilbert Scot of Thanet, according to which he could use a mill after they finished with it themselves, tends to support this conclusion. Certainly, the monks must have learned a lesson in windmill economics. It was advisable to utilize such a profitable facility once it had been built, even if its former owner violated their monopolistic franchise.

Both the layman, William Wade, and the cleric, Dean Herbert, claimed the right to do as they wished on their own land. Both were defeated by powerful ecclesiastical magnates who insisted on authorizing any innovations made in what they considered their own jurisdiction. But whereas the dean spoke eloquently about inherent rights, the knight attempted to circumvent the law to achieve his own purposes and was ultimately destroyed by it. The dean's case was settled almost instantaneously in a baronial court; the knight's case dragged on for years, and was decided in a series of royal court sessions. The Suffolk confrontation was recorded only in the chronicle of one individual. The Kent suit involved at least eleven governmental records, one personal charter, two cartulary copies of the charter, and a cartulary transcript.

Little more is known about William Wade. In 1202, 1203, and 1205—while his case was working its way through the courts and immediately afterward—he was called upon to participate in several grand assizes, so his reputation must not have been affected. Yet, no connections between him and other post-mill proprietors can be established. This is most unusual. Kent is such a small place that he must have known other windmill promoters— certainly Archbishop Hubert Walter—at least by reputation. William's apparent isolation may not have been accidental. He was possibly the least trustworthy post-mill pioneer.

Ironically, had William's final hearing been delayed a bit longer, he might have kept his windmill. In 1205 Archbishop Hubert Walter died, and after considerable turmoil Stephen Langton was chosen to succeed him. The Christ Church monks so infuriated King John by their intransigence at this time that he expelled the whole community from England in 1207. It would have been a splendid opportunity for William Wade to recapture his mill, but he did not seize the moment and King John's decree held firm.

Even though this contest was amply recorded, the public did not perceive it as important. It is surprising that Hubert Walter, who was archbishop during most of the prosecution of the case, did not become involved in it. As spiritual father of the Christ Church community, he might have been expected to show that he was concerned about the monks' welfare. Since he was also justiciar (1193–1198) and chancellor of the realm (1199–1205), he was responsible for the administration of justice. Finally, Hubert had windmills of his own—one on the Isle of Thanet and another at nearby Reculver.

The archbishop may have deliberately absented himself. The possibility of a conflict of interest probably did not concern him. It seldom stopped any Anglo-Norman administrator. Because Hubert had been quarreling with the cathedral monks for some time about his proposal to build a collegiate church, he may have felt no compulsion to assist them after they had failed to cooperate with him.

Hubert was more kindly disposed toward some other institutions, especially the Eastbridge Hospital of Saint Thomas the Martyr. He was concerned, he said, about the poverty in which its brethren were known to labor. Therefore, between 1193 and 1195 he significantly increased the hospital's endowment by granting it the tithes from two windmills (at Westhallimot, on Thanet, and at Reculver) and two water mills.[5]

Hubert was the hospital's most distinguished benefactor. He was also the most renowned of the nonroyal post-mill advocates. However, since the Reculver and Westhallimot manors had belonged to the archbishops for centuries, the wind machines there may have been pioneered by one of his predecessors. Hubert's interest in architecture was well known. According to contem-

5. See Appendix entries nos. 11 and 13.

porary observers, he was forever erecting or repairing buildings on his estates.[6] Canterbury Cathedral was then undergoing major reconstruction in the new Gothic style and one clerk in the archbishop's retinue was Master Elias of Dereham, who was later the architect, or keeper of the works, of Salisbury Cathedral.

Although he can hardly be considered an innovative pioneer, since his interest was aroused so late in the century, Hubert's experience was similar to that of many early windmill proprietors. He was born in West Dereham, in Norfolk, the second of five sons. His father was a knight of middle rank. As often happened, he was nurtured in the household of a relative. This uncle was no ordinary knight but the warrior and jurist Ranulf de Glanville, who was the realm's justiciar during the last nine years of the reign of Henry II. Living with Ranulf changed Hubert's life and ensured his governmental career. He entered the clergy, perhaps studied abroad for a short interval, returned to become a clerk for Glanville, was granted the deanship of York Minster, and in 1189 was elected bishop of Salisbury.

Hubert Walter did not devote much time to the affairs of that diocese. With his uncle, the new king, and a host of other dignitaries, including several post-mill patrons, he set out on the Third Crusade. In the Holy Land he buried the aged Ranulf and demonstrated extraordinary courage and leadership—even negotiating with the fabled Saladin. Upon his return to England he was appointed justiciar. In 1193 King Richard, though he was still being held captive in Germany, nominated him to be archbishop.

Hubert was not a learned man; complaints were made about his weak command of Latin. However, he was extremely clever and had the good sense to surround himself with bright, university-trained subordinates. His own talents were mainly adminis-

6. Gervase of Canterbury, *Opera*: 2:412. Hubert has attracted the attention of two distinguished scholars, Christopher Cheney and Charles Young. See Cheney, *Hubert Walter*; and Young, *Hubert Walter, Lord of Canterbury and Lord of England.*

trative; he preferred to mediate disputes lest they erupt into serious conflicts. He was ambitious, ostentatious, and stubborn, but no scandal, except for the brutal quashing of a London communal revolt, besmirched his reputation.

Even before he became seriously ill in the last decade of his life, Hubert sought the friendship of physicians. He is recorded as having associated with eleven of them, several on a fairly regular basis. He seems to have worn a piece of haircloth about his loins. Some might declare this mortification conventional, but such a characterization never made the garment any more comfortable for penitents. Although he was hardly a saint, Hubert Walter did serve his church and his king with industry, talent, and dedication. To many, he was a source of strength and hope in troubled times.

The archbishop's enthusiasm about windmills is probably traceable to his East Anglian heritage. He grew up in a land of post-mills and had lifelong friends who were also fascinated by windmill technology. Abbot Samson of Bury and Walter Fitz Robert, who held two knight's fees in Kent, were two such associates. Baron Walter helped Hubert establish a Premonstratensian abbey at Dereham when the future justiciar was still dean of York. Another participant in the establishment of that foundation was Hubert's cousin, Theobald of Valeins, the foremost windmill sponsor along the Norfolk coast, to whom Hubert later gave occasional judicial assignments.[7]

Not all windmill enthusiasts were his friends, however. Hubert quarreled with Philip of Burnham, sponsor of the Fincham post-mill in Suffolk. Their argument was about wardships, not windmills, and it was eventually settled in court.[8] Hubert's strong enmity toward Adam of Charing, the builder of a windmill in Romney, will be discussed later in this chapter.

---

7. B.L. Additional MS. 46353 (Dereham Abbey Cartulary), fols. 8–9.
8. On Philip see Appendix entry no. 25.

The hospital of Saint Thomas the Martyr is located on the bridge where the main thoroughfare of Canterbury crosses the River Stour. Eastbridge Hospital, as it was often called, was sometimes incorrectly said to have been established by Thomas Becket. Rather, it was founded in his honor by Edward Fitz Oldbold, a Saxon alderman of the town, between 1170 (when Becket died) and 1180. The hospital's founding was part of the second wave of hospital expansion in Kent, which continued for fifty years after Becket's death. (The first had lasted from about 1070 to 1140 and was followed by a hiatus.) The saint's nephew, Ralph, was Eastbridge's first master. He sometimes became embroiled in local Canterbury controversies and once found himself in jail as a result.

The hospital was originally established for poor and infirm pilgrims. It survived the Reformation and still carries on its work. Large parts of the twelfth-century structure are intact, and it is today one of Canterbury's major tourist attractions. The neighborhood has dramatically changed since the hospital was built. It is no longer surrounded by smelly tanneries and noisy water mills but rather by hotels, museums, and small shops.[9]

In his declaration of concern, Archbishop Hubert must have exaggerated the extent of Eastbridge's poverty. The hospital's endowment was probably quite adequate for its dozen regular inmates, and its funds were increasing. For example, before 1189 John Fitz Vivian, a provost of Canterbury and one of the witnesses to a miracle ascribed to Thomas Becket, gave one bed "on which a sick and broken down man might lie and be fed."[10] This may be the first known endowment of a single hospital bed. Lambin Fleming, a merchant who lived near the hospital, set up a fund

9. For accounts of Eastbridge, see Urry, *Canterbury*; S. Gordon Wilson, *With the Pilgrims to Canterbury and the History of the Hospital of Saint Thomas*, p. 39; Derek Hill, *Eastbridge Hospital and the Ancient Almshouses of Canterbury*.

10. Canterbury Cathedral Archives, Eastbridge Hospital Muniments, Ancient Charter A 26.

for the lighting of its infirmary.[11] As the hospital increased its daily dole to nonresidents, however, additional finances were needed. Thus, the archbishop's gift came at an opportune time.

In Canterbury there were four mills on the eastern arm of the Stour; two others were located on the river's western branch. Still others were in the vicinity. According to the terms of Hubert's gift, tithes from the Westgate mill (located on the main stream beside the city wall), another water mill at distant Herewick, and the two post-mills at Reculver and Westhallimot were to be donated to Eastbridge Hospital. Although it is not known what 10 percent of the profits from these commercial milling operations was in any given year, the tithes must have constituted a significant proportion of the hospital's annual revenue.

Eastbridge Hospital was not a passive recipient of such largess. It became active in milling and developed a special relationship with certain millers. A hospital charter dated before 1189 recorded a complicated transaction among three millers. The agreement was proudly witnessed by four other millers.[12] This was the largest assembly of such tradesmen to date. The gathering almost suggests a conference, or a meeting of a millers' guild, but no evidence of such a fraternal milling association at that time is known to exist. Contemporary documents do reveal that at least seven other millers practiced their trade in Canterbury. Such numbers suggest that the productive capacity of the six water mills must have been supplemented by other local facilities. There was a horse mill in town, and several less well known post-mills were humming away in the nearby suburbs. These structures seem to have been constructed almost simultaneously at the end of the century, and thus were probably newer than the six mills in Canterbury. Many did not belong to a manor or an institution but were commercial ventures, jointly underwritten by several part-

11. Ibid., A 15.
12. Ibid., A 46.

ners. Not surprisingly, Eastbridge Hospital became a partner in a number of these enterprises. Its brethren had developed a special interest in the economics of milling and they were quite responsive to the newest technology.

Evidently the nuns of Holy Sepulchre Convent in Ridingate, just east of Canterbury, thought that the Eastbridge Hospital authorities were more experienced in milling than they were. Sometime after 1180, the sisters concluded a detailed agreement with the hospital and gave it one-quarter of an acre upon which to build a post-mill. This is the smallest plot known to have been allocated for a windmill, but it was a significant donation for the convent, since its endowment was quite limited.[13]

The Holy Sepulchre nuns could have sought the cooperation of Christ Church Priory, but they may have been deterred by the dispute over the manor at Monkton. The monks of Saint Augustine's Abbey also knew quite a bit about milling and for a while had owned the royal water mill at Eastbridge. They had long-established ties with the Holy Sepulchre nunnery and had given it a church in Ridingate. However, Saint Augustine's Abbey controlled most of the land surrounding the convent, and the nuns may have decided to contact someone else in order to assert their independence.

The Saint Augustine monks were just as knowledgeable about wind-power technology as were the Eastbridge Hospital brethren. Spurred on by Abbot Roger (1175–1213), the monks had acquired one post-mill at Northwood, just beyond the city limits, and had built another on their manor at Chislet, not far from Reculver. The Hospital of Saint Lawrence, which the abbey operated, had a post-mill near Old Dover Road, quite close to the Holy Sepulchre installations.[14]

The nuns were sophisticated enough to know exactly what

13. See Appendix entries nos. 7, 8.
14. See Appendix entries nos. 8, 9.

they wanted. Prioress Juliana apparently spearheaded the negotiations. The contract provided that Eastbridge Hospital would pay the nuns a rent of six pence a year for the land. The convent would furnish 25 percent of the cost of building, maintaining, and repairing the mill. In turn, the nuns were to receive the same percentage of the profits and to retain the right to grind their own grain whenever they chose. One miller was to work faithfully for both institutions. The hospital would make a path from the main road to the mill (a relatively minor task in this instance) in order to attract trade.

The machine was to be built on a quarter-acre of a plot called Little Foxmold, diagonally across from the nunnery on the south side of Watling Street, which at this point was called Old Dover Road. Had one traveled that ancient thoroughfare in a northwesterly direction at that time, one would eventually have passed Wigston Parva, Wibtoft, and the post-mills of Leicestershire.

Richard of Bramford, a seneschal of Christ Church Priory, gave the Hospital of Saint Thomas the Martyr his share in a second post-mill. (His business associates are not identified.) The mill was located on a plot called Foxmold, which was also on the south side of Watling Street but separated from Little Foxmold by Ethelbert Road. Richard wanted to be sure that his soul and the souls of his family would be remembered in the prayers of the hospital patients and staff.[15]

Adam of Charing, the last Kentish windmill proprietor to be considered here, was a truly complex individual who possessed some of the undesirable characteristics of William Wade as well as certain virtues of the ecclesiastical promoters. He was in and out of trouble all his life. This is odd, since Adam seems to have practiced as an attorney. His business sometimes took him to Thanet, but there is no indication that he had any dealings with William Wade. Adam was clever but weak. He was always alert

15. See Appendix entries nos. 7, 8.

to new investment opportunities and short-term solutions but was usually oblivious to larger questions of principle. His ups and downs can be traced for half a century, but two momentous decisions, one cowardly and one generous, framed his whole career. The first was ultimately redeemed by the second.[16]

By 1164 the controversy between Thomas Becket and Henry II over the powers of church and state had become a bitter personal battle. Attempting to evade the king's mounting wrath, the archbishop set sail for the Continent from Romney, where he had a manor that controlled much of the port. Aboard ship, for some unknown reason, was Adam of Charing. Although he held the manor of Charing from the archbishop, it is almost inconceivable that he was a member of the primate's entourage seeking exile. Certainly he sat apart from him. Adam must either have had personal business overseas, or he was executing one of the minor royal commissions that he was assigned from time to time.

The ship's crew was fearful of the consequences of transporting Becket out of the country because such action was a flagrant violation of recent royal proclamations. The mariners were torn by indecision, for they wished to honor their archbishop, yet they were anxious to avoid royal notice. When well out to sea, they brought their troubles to Adam, lamenting that they had been foolish to aid the king's enemy and that they feared their kin would be severely punished as a result. Then all the sailors went together to the archbishop and declared that the wind was unfavorable. The ship hastily returned to port.

The wind can be treacherous off the coast of Romney, but that is certainly not the only reason they turned back. It is not clear whether Adam actually urged the sailors to return to port,

16. On Adam's career and that of his colleague, John Fitz Vivian, see Urry, *Canterbury*, pp. 180–81; and his "Two Notes on Guernes de Ponte-Sainte-Maxence: *Vie de Saint Thomas*"; also S. E. Rigold, "Two Kentish Hospitals Re-examined: St. Mary Ospring and SS. Stephen and Thomas, New Romney," pp. 48–49.

20. Archbishop Thomas Becket sails to France in 1164
to begin his exile. A fourteenth-century impression,
from Royal MS. 2 B VII (Queen Mary's Psalter), fol. 292;
reproduced by permission of the British Library.
An earlier attempted crossing had been thwarted by
Adam of Charing, an important windmill promoter.

or whether he merely served as a sounding board for their worries. In either case, his role was certainly not heroic. Yet, to give Adam his due, Becket was not an easy man to deal with, even in the best of times. Many others, bishops as well as barons, also deserted him when he needed them most. However, since Adam, rather than the ship's captain, was blamed for the ship's return, his intervention must have been quite emphatic.[17]

The archbishop's subsequent attempts to flee to the Continent also failed. In October 1164 he reluctantly attended the meeting of a royal council at Northampton, during which King Henry made clear his intention to ruin him and reverse his policies. Shortly thereafter, in a thin disguise, Thomas fled to France via another route and remained in exile for six years.

17. The sole account of this voyage is that of Guernes of Ponte-Sainte-Maxence, *Vie de Saint Thomas*, lines 1361–65.

Strange to say, Adam was heavily fined by the king at that time for some unmentioned offense.[18] However, it was only a temporary setback. He was soon restored to favor and made part of a royal commission charged with administering the immense revenues of the archbishopric, which, following tradition, fell under royal control during a vacancy.[19] In this capacity Adam became something of an expert on the fortification of towns and castles; he and his friend John Fitz Vivian directed the rebuilding of the walls at Canterbury and in other towns.[20]

In 1169, Archbishop Thomas thundered from abroad and excommunicated several secular and religious leaders for opposing his policies. Also excommunicated were a couple of minor Kentish figures, including Adam of Charing. Presumably, the attorney was punished because he was an unfaithful vassal and because he misused some of Canterbury's treasures in the primate's absence. Since none of his fellow commissioners, such as John Fitz Vivian, was excommunicated, Becket's grievance against Adam must have been deeply felt.[21]

The archbishop's trimphant return to England was followed by his martyrdom in late 1170. Adam was not part of that shocking tragedy. He must have made his peace with Thomas, either before his death or afterward, for he was not formally disgraced in the murder's aftermath. Nevertheless, his life must have been difficult, perhaps for decades thereafter. Not only did he endure the pricking of his conscience, but reports of his conduct at Romney were circulating in Europe; the episode was described in a

18. *Pipe Rolls, 11 Henry II*, p. 105; *12 Henry II*, p. 112; *13 Henry II*, p. 200.

19. Ibid., *13 Henry II*, pp. 201–2.

20. Ibid., *20 Henry II*, pp. 1–2; *28 Henry II*, pp. 88, 106.

21. David Douglas and George Greenaway, *English Historical Documents*, 2:750–51. For the Latin texts concerning the excommunications, see James C. Robertson and J. B. Shepherd, eds., *Materials for the History of Thomas Becket*, 3:91; 6:559, 572, 594, 603; 7:49, 99, 101, 103, 106, 110, 111.

biography of Saint Thomas written about 1175.[22] Few residents of the town of Canterbury would have been unaware of his shame.

Nevertheless, Adam continued to engage in his slightly unsavory activities. He acquired valuable properties, including a large stone house in Canterbury, by taking advantage of others' distress. He was publicly criticized for giving evasive answers while serving as an expert witness in a land inquiry. And he may have become involved in a land reclamation venture, from which he apparently later withdrew with difficulty.[23]

In about 1180, Adam made the second crucial decision of his life: he decided to found a hospital in honor of Saints Stephen and Thomas. The Norman prelate had had a special devotion to the first Christian martyr, and after his own murder, their names were often linked in church dedications and liturgical decorations. Adam deliberately built his hospital at Romney (now called New Romney), as close as possible to the place where he had betrayed the archbishop around sixteen years earlier. Obtaining the desired land had been difficult, since it was controlled by the archbishopric, which rarely alienated property, but he had a friend in Archbishop Baldwin of Canterbury and the matter was eventually resolved in Adam's favor.

The hospital was sometimes said to have been designated specifically for lepers, and sometimes for all the sick. Adam's substantial endowment reflects his wide interests. Included were more than 120 acres, income from certain other properties, and his half ownership of a windmill in the marsh at Romney. This last benefaction was earmarked to feed and clothe the chaplain. As half owner, Adam, like other promoters, had been prudently sharing the risks. The cryptic reference to his mill does not iden-

22. See footnote 17 to this chapter.
23. *Pipe Rolls, 31 Henry II*, p. 233; *32 Henry II*, p. 191; *33 Henry II*, p. 208; *34 Henry II*, p. 204. For the criticism of Adam dated about 1189, see William Stubbs, ed., *Epistolae Cantuariensis*, 2:308–9. On the land reclamation venture, see *Curia Regis Rolls*, 1:8; *Feet of Fines, Kent*, p. cxxiv; Rigold, "Two Kentish Hospitals," p. 48.

tify fellow investors or indicate whether the windmill site was also included. It probably was, since later deeds record a separate half-acre parcel.[24]

The language used in Adam's grant suggests that the mill might have been a pumping station. This would be consistent with its location in the marsh and the many drainage projects that were under way there. His hospital charters regularly stress the need to maintain the Romney sea walls. However, the term "marsh" referred both to the wetlands and to the area already reclaimed—land that was profitably used for grazing and agriculture. Since Adam's windmill produced revenue, it must have been used to grind grain.[25]

The Romney hospital soon acquired half ownership of another mill. The structure was originally jointly owned by Stephen, the son of Hamo, and William, the son of Christina, two of Adam's close associates. When Stephen made the grant about 1200, he used the ambiguous word "mill." It was probably a windmill, but one cannot be certain since the hospital ultimately owned five local mills—two windmills and three water mills. One windmill was located on the highest hill in town; the other was called Spittlemilne. Whether and when the institution acquired ownership of William Fitz Christina's half of the mill was not recorded.[26]

William Puignant witnessed the grants of both Adam and Stephen. He had once controlled the land on which the leper hospital stood, presumably leasing it from the archdiocese. His permission was therefore essential to the construction of the hospital and it was evidently given quite willingly. William was

---

24. On the Romney post-mill, see Appendix entry no. 12. On the hospital, see Rigold, "Two Kentish Hospitals"; A. F. Butcher, "The Hospital of St. Stephen and St. Thomas, New Romney: The Documentary Evidence."

25. Oxford University, Magdalene College Muniments, Ancient Charters, Romney Marsh Deeds nos. 53 (concerning two half-acre plots), 57.

26. Ibid., deed no. 37; see also Appendix entry no. 12.

known in Canterbury commercial and political circles and had a particular regard for the memory of Archbishop Anselm. He owned a post-mill near Caen, in Normandy, that he sold for six pounds Angevin in 1201. Whether Adam of Charing had stimulated William Puignant's interest in windmills, or vice versa, is anyone's guess, but their relationship might have accelerated the spread of the new technology overseas.[27]

Adam's checkered career bounced along for a quarter century after the establishment of the Romney hospital. Sometimes he gained, as when he obtained additional land in Lenham from the abbot of Saint Augustine's. This site, not far from Charing, is noteworthy because a post-mill was constructed in that area, perhaps between 1193 and 1199. Its builder was never identified.[28]

More often during this period, Adam suffered grievous loss. Worst of all, he endured the tragedy of his only son's early death. And when Archbishop Baldwin died in the Holy Land during the Third Crusade, Adam lost a powerful protector. The new primate, Hubert Walter, took a distinct dislike to him and almost immediately ousted him from his position as cathedral steward. In 1195 his former associates attempted, apparently unsuccessfully, to sue him for some of the land he had granted the Romney hospital. Their action raises the possibility that Adam may have had an imperfect title to half of the Romney post-mill, or that he may have incurred substantial debts before he donated it to the hospital.

Another loss was Archbishop Hubert's confiscation of Adam's ancestral manor at Charing, where the archbishopric already owned property. In this major attack, the archbishop also saw to

27. For information on William Puignant's post-mill, see Delisle, "Origin of Windmills," p. 404. On his English activity, see William J. Urry, "Saint Anselm and His Cult at Canterbury" (an appearance of his name about 1166); Oxford University, Magdalene College Muniments, Ancient Charters, Romney Marsh Deeds nos. 37, 53, 62; *Register of St. Augustine's Abbey, Canterbury, Commonly Called the Black Book*, 2:418, 517, 543, 590.

28. *Register of St. Augustine's Abbey*, 2:526, 483–84.

it that Adam lost certain lands on the Isle of Oxney, west of Romney, that he held from Dover Priory. The circumstances of the first action are unclear and the proceedings were noted only briefly, but in 1205, immediately after Hubert Walter's death, Adam paid the crown one hundred pounds and a palfrey (lady's saddle horse) of Gascony, a colossal penalty, to regain his manor. He claimed that the archbishop had acted arbitrarily in dispossessing him. That he won this case mattered little, for it was his final legal exercise; he followed the primate to the grave, between 1205 and 1207. His heirs assumed the responsibility of making good the title to the Oxney land.[29] The Romney hospital continued to expand. It received endowments, large and small, from Adam's friends and heirs, including a grandson who proudly identified Adam as the founder.[30]

Archbishop Hubert's campaign against Adam was probably motivated by his total contempt for the man. However, a guess can be hazarded about other reasons for the Charing dispossession. This sleepy village and its attendant manors had become important with the increase in the number of pilgrims to Becket's Canterbury shrine. For travelers coming from the west, Charing was the last overnight stop before reaching the cathedral. Perhaps Hubert Walter considered it inappropriate that the saint's enemy be known as the lord of a large manor in the area.

The archbishop had ambitious plans to make the Charing parish church of Saints Peter and Paul a shrine and tourist attraction. In the church he placed the stone that he and King Richard had brought back from the Third Crusade. They believed it to be the execution block on which John the Baptist had been beheaded according to Herod's lustful command. The stone disappeared

29. *Curia Regis Rolls*, 1:8; *Feet of Fines, Kent*, p. cxxiv; *Pipe Rolls*, 7 *John*, p. 117; 8 *John*, p. 51; 9 *John*, p. 34; 10 *John*, p. 101. See also Rigold, "Two Kentish Hospitals," p. 48.

30. Oxford University, Magdalene College Muniments, Ancient Charters, Romney Marsh Deed no. 53.

during the Reformation, but for more than three centuries it brought fame and wealth to Charing.[31]

Fiscal management also influenced the archbishop's plans. His general policy was to return all leased manors to direct control, especially because inflation had made low rents more unprofitable than ever. Charing was one of the last to be affected. Its twelfth-century payments to the archdiocese were less than they had been in 1066, 1086, and 1090.[32] Hubert Walter was determined that Adam should no longer profit at the church's expense.

Perhaps death came as a welcome relief for Adam. The available documents do not reveal his feelings, but one senses that old, alone, and financially troubled, he may have finally tired of life. He had always been a fighter—not for great causes but for his own advantage. The battles had taken their toll. Only the haunting memory of one windy voyage was ever fresh.

The pretentious quiet of his great mansion must have allowed Adam ample opportunity for reflection. If he ever surveyed his years and days, he probably concluded that his career had had too many peaks and valleys to sustain the twelfth-century analogy that life was a turn on the wheel of fortune—a slow rise, a brief stop at the height, and a precipitous fall. Didactic preachers might compare Adam to a sinner who sowed the wind and reaped a whirlwind, but the Charing lord may have viewed himself as someone who had survived Fortune's fickle blasts. Adam's path was narrow and twisted, seeking accommodation, pressing minor distinctions, and ultimately bargaining for salvation. It followed the trend of the times, but it led to a hollow in the heart.

31. *About Charing: Town and Parish*, pp. 14–16.
32. Francis R. H. Du Boulay, *The Lordship of Canterbury*, pp. 200–205.

# The Outer End of the Beam

*T*he post-mill boom that occurred in Leicestershire, on the manors of Walter Fitz Robert, within the liberty of Bury Saint Edmunds, and throughout the holy land of Kent spread rapidly to other shires. Sussex promoters boasted of their facilities and Norfolk marshmen quietly built more windmills than any other group. Elsewhere, questions of mere property rights became significant issues of principle.

At the same time that installations were embracing the Channel air in Kent, an unidentified but ambitious Cambridge knight built his own windmill on land that had rendered tithes to the church. Presumably, this means the land was once cultivated. After erecting the post-mill, he stoutly refused to pay any ecclesiastical tax on the land or on the profits of the new mill.[1]

The available records do not indicate the justification for his action, which clearly flouted established custom. Although his refusal lacks the drama of Dean Herbert's protest, the issue was equally important. Debate about his action probably continued

---

1. For more information about this windmill, see Appendix entry no. 4.

for several years. One source suggests that two knights were involved—one who constructed the mill, and another who refused to pay the tithe. The parish where the mill stood is identified only as N——.

Scholars have uncovered some facts, however. The absentee rector of the church was a wealthy cleric named Burchard, who was the treasurer of York Minster and an archdeacon of both the Durham diocese and the East Riding of the York archdiocese. He was a nephew (or maybe a son) of Bishop Hugh de Puiset of Durham. Burchard was most active in ecclesiastical affairs between 1170 and 1196 and would certainly have known of the Beeford, Newcastle, and Weedley windmills. In fact, he can be found witnessing documents along with other windmill sponsors. Burchard was no man to be trifled with and he ultimately brought the problem of the knight's mill directly to the attention of Pope Celestine III (1191–1198), complaining that the knight, at the peril of his soul, had refused to pay tithes on the profits of his windmill.[2]

The obligation to pay tithes was familiar to all Christians. Many lords, such as Reginald Arsic and Archbishop Hubert, chose to donate their windmill tithes in perpetuity to a religious corporation of their choice. Other landowners were less far-sighted, or less generous. As a result, there were seemingly endless minor disputes about who should render and who should receive payments. Every producer wanted an exemption and every ecclesiastical association was anxious to receive the tithes of local businesses.

2. Ibid. See also Philip Jaffé, ed., *Regesta Pontificum Romanorum ab Condita Ecclesia ad Annum post Christum datum MCXCVIII*, document no. 17620; *Decretalium D. Gregorii IX et Privilegi Gregorii XIII*, p. 484 (bk. 3, title 30, document no. 23); Mary Cheney, "Decretal of Pope Celestine III," is an example of superior historical research. On Burchard, see D. S. Boutflower, *Fasti Dunelmenses: A Record of the Beneficed Clergy of the Diocese of Durham Down to the Dissolution of the Monastic and Collegiate Churches*, p. 21; Arthur M. Oliver, ed., *Early Deeds Relating to Newcastle-upon-Tyne*, p. 198; William Farrar and Charles T. Clay, *Early Yorkshire Charters*, 9:143, charter 71 (1189–1192), which Burchard attested with Robert of Peasenhall, owner of the Newcastle windmill, and Thomas of Amundeville, brother of the donor of the Swineshead windmill.

The tax on the profits of water mills was justified largely by an appeal to laudable custom. However, some water mill owners tried to evade the tithe by arguing that since tithes were not collected on living animals (which was true), and since millponds were really fishponds, anything in those waters, whether living fish or a mechanized facility, was immune to the levy. This was a specious rationalization, but a few disputes about water mill tithes were still unsettled in 1226.[3] Some windmill owners claimed that they were dealing with a wholly new circumstance that was not covered by existing regulations and that they therefore deserved exemption.

Celestine III was not about to accept that line of reasoning. This aged but stubborn pontiff was unusually interested in English affairs. In response to Burchard's complaint, he invoked the general principle that every faithful person was bound to contribute tithes on the profits of all that he had lawfully acquired. The pope therefore ordered the windmill tithes in N—— to be paid regularly and without further appeal. He directed the archdeacon of Ely (Richard Barre) and the abbot of Ramsey to ensure their collection. The assignment of these prelates suggests that the parish was in Cambridgeshire. Clearly, the pope did not believe the new technology merited exemption from ecclesiastical regulations.

Ramsey Abbey sponsored more than a dozen windmills in the early thirteenth century, but these mills cannot be dated in the twelfth.[4] The brother of Richard Barre had witnessed a post-mill transaction at Wigston Magna in Leicestershire.[5]

It is unlikely that either Archdeacon Burchard or the knight considered this papal ruling a precedent. At the same time that

3. *Godstow Cartulary*, 1:lxxvii.

4. At least twenty post-mills can be identified as existing by 1250; *Ramsey Cartulary*, 1:103, 268–81, 333, 354, 397; 2:22, 225, 229, 313, 316.

5. On Richard Barre, see Diana E. Greenway, *John Le Neve's Fasti Ecclesiae Anglicanae, 1066–1300*, 2:50; Ralph V. Turner, *The English Judiciary in the Age of Glanville and Bracton, c. 1176–1239*, pp. 89–104. On his brother Hugh, see Chapter 3, footnote 41.

the archdeacon submitted his complaint about windmill tithes he made a separate, quite routine plea to the pope concerning a man who had refused to offer his tenth of hay, which had been grown on land that was normally tithed. Celestine resolved this problem by issuing a similar ruling.[6]

England generated an inordinate proportion of all the litigation heard in Rome; consequently it was also responsible for a large number of the recorded papal decrees. A national fascination with the procedures of justice was certainly one reason. English kings tried to discourage the seeking of redress beyond royal authority, but the appeals continued to increase, especially after the martyrdom of Thomas Becket. The fact that the Vatican's first recorded reference to wind technology appears in a document pertaining to an English appeal is another confirmation of the theory that post-mill development originated in England.

It is surprising that the tithing issue had not received papal attention sooner. By the time Celestine issued his ruling, the Wigston Parva windmill had been spinning for more than fifty years and several monasteries were already the happy recipients of post-mill tithes. Although it seems unlikely that Archdeacon Burchard's intractable knight was the first windmill builder to try to avoid paying a tithe, he may have been one of the first to argue that tithing of the profits of the new wind machine was inappropriate.

Perhaps earlier, less complex disputes had been handled locally or were denied appeal because they raised no major issue. The absence of King Richard (during the Third Crusade and his subsequent detainment in Austria) undoubtedly encouraged some barons to break the law and prompted some clergymen to pass their more complicated problems on to Rome. The increase in the number of trained canon lawyers throughout Christendom was a mixed blessing. It ensured more thorough treatment of important

6. Jaffé, *Regesta Pontificum Romanorum*, document no. 17621.

concerns, but it also probably multiplied the questions lawyers wanted answered.

Why the decision of Pope Celestine concerning windmill tithes later attracted attention is not known, but in 1234, when hundreds of papal decrees were being collected to form the basis of a new, authoritative compendium of canon law, Celestine's two rulings in response to Burchard's complaints about post-mill and about hay tithes were incorporated as definitive statements. Henceforth, they were considered precedent-setting decrees.

Some of Celestine's other injunctions were also preserved by local religious authorities. At about the time the decretal collection was being compiled, a scribe at Wymondham Priory in Norfolk copied the charters of his convent into one large volume that included records pertaining to five windmills. The cartulary began with transcripts of several papal letters. One, from Celestine III to Saint Albans Abbey (Wymondham's motherhouse), concerned tithes. That abbey was forever consulting the papacy on some issue; it obtained more than one hundred papal replies between 1159 and 1197. In his letter, Celestine referred in passing to the tithes from "new mills." The scribe apparently thought the message had some pertinence for his house and pessimistically placed it beside a papal letter warning monks not to associate too closely with laymen.[7]

One Englishman whom the pope respected for his legal expertise was Bishop Seffrid II of Chichester (1180–1204).[8] This comparatively obscure prelate was an avid advocate of windmill construction in his own diocese, and he seems to have taken more personal delight in the new technology than any of his contem-

7. B.L. Cotton MS. Titus C VIII (Wymondham Priory Cartulary), fols. 14–14v, issued at the Lateran in 1193; printed in Walther Holtzmann, *Papsturkunden in England*, 3:543–44, 553–54. On papal bulls to Saint Albans, see Christopher R. Cheney, *From Becket to Langton: English Church Government, 1170–1213*, p. 85.

8. Cheney, *From Becket to Langton*, p. 29. The pope similarly respected Master Godfrey de Lucy, bishop of Winchester.

poraries. He boasted that he built windmills at his own expense, enthusiastically described some of their unique features and took steps to make it easier for customers to reach them.[9] Seffrid had extensive knowledge of East Anglian affairs, having served there as a royal justice. He knew other wind power promoters, such as the earls of Arundel, who were also deeply involved in Sussex and East Anglia. Thus, the bishop was unusually well qualified to counsel the pope about windmill matters.

Seffrid II may have been related to an earlier bishop of Chichester, Seffrid I (1125–1145). Their uncommon first name suggests some type of kinship. The younger man was active in the diocese for a long time, from at least 1154. Before his consecration he was called Master Seffrid, probably because he had completed advanced university study and had a license to teach. He probably obtained his legal training at the University of Bologna. A twelfth-century necrology from a church in that city includes prayers for the mother of a Seffrid the Englishman.[10]

As was so frequently true, diligent study opened the doors to ecclesiastical preferment. In 1172 Seffrid was made the archdeacon of Chichester and later was elected dean of the cathedral chapter. During this time he was given a number of benefices, including the rectorship of Bloxham Church in Oxford, a post formerly held by Bishop Robert II of Lincoln and by the royal physician, Master Ralph de Beaumont.

The excavation of his tomb reveals that Seffrid was five feet eight inches tall. He was buried with a plain silver chalice and paten, perhaps possessions dating from his early priesthood. A gold episcopal ring set with a sapphire and five emeralds encircled

9. See Appendix entries nos. 51, 53.

10. On the younger Seffrid's career, see Henry Mayr-Harting, *The Bishops of Chichester, 1075–1207*, pp. 14–16. He was friendly with physicians, especially a Master Hugh; see C. H. Talbot and E. A. Hammond, *The Medical Practitioners in Medieval England: A Biographical Register*, p. 90.

one finger, and in his hands had been placed an elaborately carved ivory crosier—both indicators of his clerical success.

The bishop was one of the most prodigious builders in the history of the diocese, and he liked to talk about his architectural achievements. Seffrid strengthened the walls of the episcopal castle at Amberley, enlarged the Church of Saint Andrew in Bishopstone, and rebuilt the portions of Chichester Cathedral that were destroyed by fire in 1187. In reminding everyone that he erected windmills and definitely paid for them himself, he seems to have been forestalling complaints about his construction program.

Seffrid's interest in windmills may have been stimulated by William de Vescy, an important Northumberland baron and a member of one of Britain's most learned families. William's father, Eustace Fitz John, was a counselor to several kings and an active philanthropist. William had dozens of knights as his vassals and held properties throughout the realm. Some of these tenements, such as Weedley in Yorkshire, North Ormesby in Lincolnshire, and scattered properties in Suffolk, were near windmills. He also knew a number of windmill promoters and his son was a close friend of Walter Fitz Robert's heir.

The fact that William planned to build a windmill in Sussex and was willing to pay dearly to do so suggests that he might have erected other post-mills on his own northern barony. At some unspecified time before 1183 the dean (who was not mentioned by name but could have been Seffrid) and canons of Chichester Cathedral entered into a contract with William de Vescy, which provided that he could rent their mill site on Ecclesdon hilltop in Ferring and build a mill there at his own expense. He was to pay the chapter one silver mark a year, half at midsummer and half at Christmas. This fairly high rent suggests that both the canons and the baron must have expected the mill to do well.

The contract, like many early agreements, did not specify a

windmill. No other installation on that hilltop would seem practical, however. Use of the term "mill site" might indicate that a mill had previously been located there.

William never built his mill, however. He sensed that the end of his life was near, and he quietly retired to his northern homeland to spend his final days as a Premonstratensian canon at Alnwick Abbey. He died in 1183. After a brief interval, Bishop Seffrid took over the project and applied his own immense energies to it. Later he proudly described his windmill in unusual detail.

Sometime before 1197, Seffrid gave this same post-mill to Thomas of Ferring, his household steward, as a reward for his services to the cathedral and to the bishop. His charter grant included the two acres next to the road that leads to the mill and "the breadth of an acre all around the outside of the outer end of the beam by which the mill is turned round." The bishop further announced that he had erected the windmill at his own expense and that he would allow free access across his land to anyone who wished to use the mill. For the grant of the mill and the land, Thomas agreed to pay the Chichester Cathedral chapter one pound of pepper each year, a high rental.

Since he had built the windmill himself, Seffrid was anxious to attract customers. He understood how important it was that the mill be near a road and that the road be open to all.

Seffrid's reference to the exterior beam is the earliest verbal description of a post-mill. It verifies that the twelfth-century structures were accurately depicted in later manuscript illustrations. Like several of those miniatures, this description emphasized the tail pole. However, whereas the bishop concentrated on the acreage beneath the circling beam, the painters whimsically depicted dogs or sacks of grain riding on the pole. The Ecclesdon windmill, like the installations at Cleydon, Hempnall, Hickling, and Swineshead, was placed in an open field, thus eliminating many transportation costs and allowing utilization of high ground and unobstructed air.

21. Continental artists began to depict windmills about 1270,
at least a decade after English painters sketched them.
This charming scene was painted by Jehan de Grise of Bruges
between 1338 and 1344. From Bodleian Library, MS. 264
(*The Romance of Alexander*), fol. 81; reproduced by permission
of Oxford University. Another, slightly damaged
post-mill appears on fol. 49 of the same manuscript.

Thomas the steward was either extraordinarily generous, exceptionally unimaginative, or perhaps close to death, for he soon gave the Ecclesdon windmill to the new Augustinian priory at Tortington. Almost immediately, those canons sold the mill to the Chichester Cathedral chapter for twenty-five marks. All this happened between 1183 and 1197. (Interestingly, the amount was the same as that paid by the Canterbury monks to King John to ensure a favorable judgment.) Again, the terms "mill" and "windmill" were used interchangeably.

The Ecclesdon post-mill was evidently kept in good repair. For years it was regularly mentioned in both episcopal and papal correspondence, along with the diocese's other windmills at Amberley and Bishopstone.

The post-mill at Amberley was another of Bishop Seffrid's creations. In 1185 the bishop, who then had been in office only a few years, boasted that he had given his cathedral canons a windmill at Amberley "which I had first built," as well as some tithes of hay from the associated episcopal manor.

Assessments of the bishop's role in windmill construction became increasingly inaccurate over the years. In 1375, the author of a short history of the bishops of Chichester claimed that Seffrid had given the diocese a windmill at Bishopstone, but he forgot to mention those at Amberley and Ecclesdon. This writer was doubly incorrect. Seffrid gave the chapter a water mill, not a windmill, at Bishopstone. At the same time he also gave the tithes of Liffielm the miller. A post-mill was evidently not erected on Werdon Down in Bishopstone until the term of one of Seffrid's successors, Bishop Ranulf (1217–1222). In his charter, that prelate copied Seffrid's graphic language about the land beyond the outer end of the beam by which the windmill is turned round, except that Ranulf gave only rods whereas Seffrid had given acres.

The windmill and water mill at Bishopstone had a convoluted subsequent history. Seffrid paid Reginald of London (probably the same Reginald who appeared at Saint Bartholomew's Hospital, London) five marks for the water mill and then deducted one mark for its annual yield. Years later Reginald's heir claimed the mill as his own. His claim was denied, but the chapter offered him both mills for a lump payment of four marks and a similar yearly rent. The new lessee then sublet the two mills to an individual who somehow managed to sell them back to the Chichester chapter for sixteen marks. In 1243 the mills' annual profits totaled four pounds, so the chapter had evidently been wise to repurchase the two facilities.

The transmission of information within Bishop Seffrid's circle of friends, which included Pope Celestine, the earls of Arundel, and William de Vescy, illustrates how knowledge of post-mill technology spread. Most of the wind-power pioneers had similar personal networks, but they were especially prevalent in East Anglia, the most densely populated part of the kingdom and the region with the strongest concentration of windmills. The area's powerful air currents, listless rivers, and abundant fertile soil help to explain why so many post-mills rose there, but the enterprising spirit of its inhabitants was probably equally important.

East Anglia was the location of at least three clusters of post-mills that private owners donated to major religious institutions. The experience of the Abbey of Bury Saint Edmunds has already been discussed. Two other fortunate recipients of massed wind engines were Wymondham Priory and Hickling Priory, both in Norfolk. Their benefactors, the earls of Arundel and the marshmen of the Norfolk Broads, indicate the range of the new machine's appeal.

Wymondham, a Breckland market town, is just up the road from Hempnall. A wealthy Benedictine monastery and the Norfolk residence of the earls of Sussex were located there, as were at least two early post-mills. The de Albini earls were deeply involved in Norfolk affairs, even though their patrimonial seat, Arundel Castle, was not far from Chichester. William the Conqueror had granted the manor of Wymondham, and its first lord and his descendants had lovingly improved it. Wymondham is still an unusually attractive settlement, whose name is locally pronounced "Windham."

Five men, all named William de Albini, ruled this fief throughout the twelfth and early thirteenth centuries. William III (1176–1193)—or possibly his father (the documents are unclear)—gave Wymondham Priory a few windmills. His brother, Hugh, gave the priory another windmill at Betwick. Vassals and family friends donated other mills at Snettisham, far up the northern coast, and at Flockthorpe. The facilities in Wymondham were intended to be auxiliary stations and were not to be used when the local water mill was operating. The Snettisham windmill gift carried the stipulation that thirteen poor people were to be fed from the mill's revenues once a year.[11]

Although not all these grants, or the structures themselves,

11. See Appendix entries nos. 23, 26, 32, 34, 35.

can be dated precisely, the mills were undoubtedly part of the construction boom that produced windmills at Burlingham, Fincham, Hempnall, Herringby, Hickling, Palling, Rollesby, Waxham, and Wrogland in Norfolk and at Dunwich, Elveden, Haberdun, Hienhel, Risby, Timworth, Tostock, and Willingham in Suffolk. Some of these sites are difficult to identify today, but because East Anglia is relatively small, the news of a good invention traveled quickly, once a few individuals had become acquainted with its mysteries. Perhaps the clever practice of signaling from height to height with mill sails had already begun. Simple messages would later be sent great distances by millers who ritually silhouetted their crossed sheets at certain predefined angles to the horizon.

The earls of Arundel were not typical Norfolk windmill sponsors. Most promoters were small landowners who would now be described as middle-class individuals. East Anglian, particularly Norfolk, villages were friendly settlements composed mainly of prosperous free peasants rather than bonded serfs. These villages were rarely unified into a single barony; some were ruled by five or six lords. This meant that manorial authority was weaker there than in other parts of the country. Small freeholders transacted their own business and issued their own writs. Many Norfolk windmill charters are hard to date precisely because the donors were obscure independent farmers, not well-known aristocrats.

The Norfolk Broads, the marshy coastal stretch around Hickling, proved especially inviting to early millwrights. Here meandering rivers flooded old peat excavations and formed large shallow lagoons. The vegetation is lush in summer and many species of wildlife nestle in the reed grass, but in winter cold winds blow from the North Sea. Immense effort has been expended to dike and drain every foot of dry land, but the isolated hamlets are primarily linked by boat. Today, this flat area is still dotted with windmills and is a summer playground for anglers, naturalists, and weekend sailors.

22. In this early-fourteenth-century scene, the devil takes
Christ up a high mountain and tempts him with the joys of
this life. The post-mill and four types of buildings may
signify worldly wealth. The undercarriage of the mill is similar
to those of the mills in the contemporary Smithfield Priory
Decretals (see Figures 23, 24). From Additional MS. 47682 (*The
Holkham Bible Picture Book*), fol. 19v;
reproduced by permission of the British Library.

In Anvegin times, the Norfolk Broads were known for peat,
fish, and probably salt. Arable plots were small and valuable.
Neighbors often formed groups and established cooperatively operated mills. Hickling owners, unlike mill owners elsewhere, did
not forbid the establishment of multiple facilities. There were at

least six early post-mills within a few miles' radius of that hamlet.[12]

The absence of regulations against competition bespeaks prosperity and self-confidence, but there may have been an additional, more pragmatic reason. Some of these early post-mills may have been pump mills used for land reclamation, rather than flour mills. If they were, such a policy would have been in everyone's best interest. Most of the windmills still standing in the area were obviously once used for marsh drainage.

The first identifiable windmills in the Norfolk Broads were all associated with the Augustinian priory of Hickling, which was established in 1185. The founder, a knight named Theobald of Valeins II, enlisted a score of local benefactors, including many of quite average social standing. Most of them gave the monastery a series of modest gifts rather than one large grant.

Theobald succeeded his father, Robert of Valeins, in 1178. The family's ancestral lands consisted of five and one-half fees centered at Parham in Suffolk, properties in Hickling, and scattered holdings in Cambridgeshire and Yorkshire. One of Robert's sisters, Bertha, married Ranulf de Glanville, who was justiciar during the last nine years of Henry II's reign and probable author of the *Treatise on the Laws and Customs of the Kingdom of England*. He, too, held lands near Hickling and Parham. Theobald's other aunt, Matilda, married Hervey Walter, a leader of the Second Crusade's separate expedition to Portugal in 1147. One of their children, Hubert, entered Ranulf de Glanville's service and later became justiciar and archbishop of Canterbury, as well as a sponsor of two windmills. A number of men from the Hickling area, such as Reiner of Waxham and Henry Flegg, also joined Glanville's entourage. The Glanvilles, like the Valeins, were a large and complex family. Agnes de Glanville was part owner of two

12. Ibid. See Appendix entries nos. 24, 28, 29, 30, 31, 33.

Newcastle windmills. Ranulf de Glanville died in Palestine in 1190, during the Third Crusade.

Like his celebrated relatives, Theobald generously supported Augustinian and Premonstratensian monasteries. His uncle, Ranulf, established Butley Priory, near Parham, for the Augustinian canons in 1171 and founded Leiston Priory for the Premonstratensians in 1183. Theobald established Hickling Priory with twelve acres in 1185. Three years later, Hubert Walter founded a Premonstratensian house at West Dereham, in Norfolk, on land that was ultimately given by Walter Fitz Robert. Theobald attested the grant. In about 1189 he helped support Coverham Abbey, in Yorkshire, a Premonstratensian monastery founded by Ranulf's daughter. In 1195 he endowed the convent at Campsey Ash, near Parham, a nunnery built by his two sisters. On another occasion he gave five acres to Sibton Abbey, a Cistercian house.

Theobald knew Walter Fitz Robert quite well. In status and wealth the knight of Hickling was dwarfed by the lord of Dunmow, but they both sponsored mills, endowed monasteries, and served the government. Theobald's political career was quite localized. In 1194 his first cousin, the new archbishop and justiciar, appointed him a royal justice. In 1200 he became sheriff of Norfolk and Suffolk, an important and lucrative position. However, for some reason, he retained it for only a few months. Theobald died sometime before 1209.

Hickling Priory attracted a major share of Theobald's attention after 1185. Among his gifts was the site for a post-mill—a plot he described as 100 feet long by 100 feet wide and some 12 feet from the nearest road. This was considerably smaller than the one-acre sites several other English promoters gave their charities, but arable ground was extremely valuable in the marshy Broadlands. Proximity to a road was also mentioned in other transactions. Theobald must have been aware that potential customers would require easy access to a mill. No one wanted to pay a surcharge for crossing enclosed property in order to grind grain.

Theobald gave the land for a windmill, but evidently not the machine. Some smaller landholders did just the opposite. They gave Hickling Priory a windmill and their half of the land on which it stood, and then helped the priory to make arrangements with regard to the other half. One donor gave the canons her windmill in Rollesby and described its location in detail, including its distance from the road, the bridge, and the neighboring settlements. Possibly she found such directions necessary because this was not her only post-mill. Despite such generosity, the canons had difficulty collecting tithes on windmills they did not own—another indication of the numerous quartered sails in the Norfolk Broads.

All the writs in the Hickling Priory Cartulary have a slightly eccentric, overly specific format, as if they had been prepared by copyists who were told what they were supposed to do but had not had much practice at it. As in many charters, the benefactors gave the mills for the salvation of their souls and those of their ancestors, living family members, descendants, and associates. Such commemorations sometimes appeared more than once in the Hickling documents—an instance of unusual and seemingly unnecessary piety.

The declarations of first-generation patrons were usually much more pious than those of their heirs. Another quaint feature of the Hickling conveyances was that the oldest deeds often merely referred to mills, as had the first Wigston Parva charter. Only in later documents were these structures specifically identified as windmills, and even then an atypical word, *venticum*, was often used.

Each of the villages in the Norfolk Broads could have comprised only two hundred residents, or less. Yet dozens of men and women rushed to endow Hickling Priory; several donated windmills. Such generosity is eloquent testimony to their religious faith and to their enthusiasm for wind mechanics. If small villages had so many post-mills, then large towns, with their greater need for ground meal, doubtlessly had more. Bulky products such as milled flour rarely traveled far in the Middle Ages, so numerous windmills and water mills must have operated near such populous centers as London, Bristol, and York. However, we know only that windmills existed at Bury, Canterbury, and Newcastle (where there were two), and that a post-mill probably operated at Clerkenwell, outside London.

There are indications that other windmills did operate near densely populated areas. Nicholas de Sigillo (d. 1187), a Hampshire landowner, prominent government official, shipping magnate, hospital patron, friend of windmill promoters, Huntingdon archdeacon, and canon of Saint Paul's Cathedral may have sponsored two post-mills just outside Southampton. The installations were on lands he had once held but can be dated only to the early thirteenth century. At that time, two other windmills operated at Grimsby and at Warboys, towns where Nicholas also held tenements. The coincidence of his property and the identification of four post-mills suggests that the machines probably existed long before they were first recorded.[13]

13. For information on the Southampton mills, see *Saint Denys Cartulary*, pp. lxxv–lxxix; 1:30–31, charter 49; 2:150, charter 249; 2:169, charter 284. On the Grimsby windmill, see *Pipe Rolls*, 5 *Henry II*, p. 64; 6 *Henry II*, p. 45; 7 *Henry II*, p. 15; and 8 *Henry II*, p. 17. See also Edward Gillett, *A History of Grimsby*, p. 14. On the Warboys windmill, see *Ramsey Cartulary*, 2:170–74, 272; *Ramsey Chronicle*, pp. 301–2. The donor of the land in Grimsby (perhaps including the post-mill) was William Fitz Ralph, evidently the same man who also gave an early windmill to the Spalding Priory; B.L. Harley MS. 742, fol. 10. It is not clear whether this William is the well-known government official, William Fitz Ralph. Grimsby Priory also sponsored construction of a windmill at Manby, Lincolnshire, before 1200; see the Appendix entry no. 18. Nicholas de Sigillo leased the manor of Ardleigh in Hertfordshire from the canons of Saint Paul's Cathedral; see *Domesday of Saint Paul's*, pp. 22–25, 27, 111; Guildhall Museum, Saint Paul's Cathedral Muniments, Ancient Deed Press A, Box 40, no. 1412. Between 1192 and 1222, Richard of Stapelford built a windmill there; see *Domesday of Saint Paul's*, p. 21; Gibbs, *Early Charters*, p. 59, charter 82.

Certainly, windmill promoters had ample incentive to build both near large towns and in tiny hamlets. Tilting at windmills became chivalrous idealistic sport after the days of Cervantes, but there was nothing quixotic about the first post-mill owners. They were generous to local charities, but they were also shrewd competitors; at least nine individuals or institutions controlled multiple installations.

# Pushing Round the Tail Pole

*E*conomics, politics, and knowledge inevitably affected post-mill development. Detecting such influences requires going over some familiar ground again, but only briefly. The real attractions are the activities of people.

Peace shed its blessings on England in the opening decades of the twelfth century, probably the country's most prosperous era since the early Roman occupation. One result was the wondrous post-mill built on the land of William the Almoner before 1137. Thereafter windmills entered the records in mounting numbers, a little hesitantly at first, but then like steps on a bar graph, block formations symbolizing quantitative leaps in a progress report. A second post-mill is thought to have appeared in the 1140s; a third is definitely known to have existed in the 1150s. Three more functioned by the 1160s; another one by the 1170s; nine more by the 1180s; and twenty more by the 1190s. Another twenty were constructed by 1200, but they cannot be assigned to a specific decade. Although more undoubtedly existed, and many of the identifiable mills existed long before they were recorded, this assembly of fifty-six examples clearly suggests that the invention did not become truly popular until the 1180s.

Some factors encouraging the windmill's birth were discussed in Chapter 3. Now, two further questions come to mind. Why did the machine spread so slowly between 1137 and 1180? Why did it proliferate after 1180? Like the fleeting moments of sunrise, before fire light consumes the dark, the years between 1137 and 1180 blazed with contrast. On the political level the anarchy of King Stephen's reign was followed by the early achievements of Henry II and then, after 1170, by the tragic quarrels between that king and his rebellious sons. On the religious level, the numbers of monks and nuns, of monasteries and hospitals, soared through the 1150s, but the Church suffered during the bitter controversy between Henry II and Thomas Becket. By the time of Richard and John, basic economic forces were more significant for windmill distribution than the activities of monarchs or clergymen.

Despite crop yields that would be considered low by any modern standard, twelfth-century England was a grain-rich country. Wheat, barley, oats, and rye were cultivated almost everywhere and people consumed prodigious amounts of dark bread and ale. An internal market system had developed early in the century and substantial amounts of grain were soon being exported (sometimes illegally)—especially to Scandinavia.[1] Huge late-century barns offer eloquent testimony to the productivity of the

1. The earliest attempt to regulate grain exports seems to have been that made in 1176–1177; *Pipe Rolls, 23 Henry II*, p. 136. For a valuable discussion of agricultural economics, see Gras, *English Corn Market*, esp. p. 110; also M. M. Postan, *The Medieval Economy and Society: An Economic History of Britain, 1100–1500*, pp. 139, 231. On crop yields, see W. Harwood Long, "The Low Yields of Corn in Medieval England"; B. M. S. Campbell, "Agricultural Progress in Medieval England: Some Evidence from Eastern Norfolk"; Postan, *Medieval Economy and Society*, p. 51; Georges Duby, "Medieval Agriculture," in Carlo M. Cipolla, ed., *The Fontana Economic History of Europe*, 1:179, 191–95; Titow, *English Rural Society, 1200–1350*, pp. 80–81.

land, which fortunately increased faster than the burgeoning population.

Cereal crops were grown on farms of several types. The most common nonsubsistence units were the manors, large and small. A primitive communications system, regional variations, and personal preferences prevented the establishment of any nationwide manorial policy. However, two factors dominated most agricultural decisions: an alternation between personal overlord management and leasehold arrangements, and a long-term highly successful effort to bring additional acres under the plow through drainage and forest clearance.

In general, manors were "farmed out" to leaseholders when times were uncertain. Such contracts provided lords with a fixed rental and allowed them to delegate management responsibility. Simultaneously, peasant labor services were often commuted for money. Most lords wanted payment in cash rather than payment in kind (that is, produce). However, if economic conditions improved, or if prices rose dramatically, as they did in the late twelfth century, cash contracts could prove disastrous for overlords. Tenants wanted secure arrangements; many wanted contracts extending beyond their life-span. Lords sought the guaranteed income that such contracts would provide but feared that a change in economic conditions might leave them at a disadvantage.[2]

People had not yet defined inflation, but they still had to endure its effects. By the late twelfth century most lords were trying to recover the losses they had sustained as a result of unfavorable contracts. Abbot Samson of Bury Saint Edmunds, for

---

2. There are many fine studies of twelfth-century economic and (or) agrarian conditions; see especially J. L. Bolton, *The Medieval English Economy, 1150–1500*; R. H. Hilton, *Peasants, Knights, and Heretics*; Edward Miller and John Hatcher, *Medieval England: Rural Society and Economic Change*; Postan, *Medieval Economy and Society*; Titow, *English Rural Society*; and Cipolla, *Fontana Economic History*, vol. 1.

example, upon assuming office found the monastery deeply in debt and concluded that many of its manors were unprofitable. He therefore canceled or renegotiated contracts right and left. Many other lords envied his unscrupulous success.

Thousands of small landowners did not have the option of leasing, for their total holdings were never large enough to subdivide. These men were necessarily self-reliant, for to pursue any other course would inevitably lead to failure. Their holdings were concentrated in the regions where most of the early windmills were located—in southern Leicestershire, East Anglia, and eastern Kent. Those areas had traditionally been inhabited by free peasants; numerous knights of modest means had also taken residence there. In Wigston Magna and along the Norfolk coast, near Hickling, free peasants and provincial knights cooperatively sponsored new mills. The joint ventures in the vicinity of urban Canterbury were similar, except that ecclesiastical establishments, both large and small, were more likely to be the post-mill proprietors.

Increasingly, landlords decided not merely to react to change but to plan for it systematically. As early as Henry I's reign (1100–1135), inquiries were regularly made of royal estates, not only to check revenues but also to take inventory of stock and to examine resources. Monasteries made similar studies. By the end of the twelfth century, the lands of the bishops of Winchester were being subjected to an annual audit. Formal treatises on accounting and estate management appeared in the mid-thirteenth century.[3] Such training manuals dating to the twelfth century have not yet been found. However, in 1179 the royal clerk, Rich-

---

3. *Leges Henrici Primi*, ed. L. J. Downer, esp. p. 175. For an excellent analysis of twelfth-century estate management practices, see the introduction to Marjorie Chibnall, *Charters and Custumals of the Abbey of Holy Trinity, Caen*. On estate management treatises, see Dorothea Oschinsky's introduction to *Walter of Henley and Other Treatises on Estate Management and Accounting*; see also Titow, *English Rural Society*; Postan, *Medieval Economy and Society*, pp. 96–99; Duby, "Medieval Agriculture," pp. 204–11.

ard Fitz Nigel, completed a lengthy analysis called the *Dialogue of the Exchequer*. He explained in detail how the financial machinery of the government functioned and even discussed why churchmen should be interested in seemingly trivial secular affairs.

Some contemporary windmill proprietors established management practices of their own. For example, many of Walter Fitz Robert's prolific writs were prepared by a scribe who dutifully attested them himself. A similar secretary drafted several deeds for the Eastbridge Hospital brethren. The bishops and abbots already had special clerical departments.

In contrast to the ebb and flow of estate management, the continuous surge of land reclamation was one of the most significant developments in Western Europe during the twelfth century. In England alone countless acres were reclaimed from fen and forest. This incredible effort, though largely unrecorded, had been under way in some areas, such as Romney Marsh, since Roman times.[4]

The coincidence between areas drained and windmills erected is astonishing. Major reclamation projects were carried on in the Humberland marshes of Yorkshire, in Holderness, in the Cambridge fens, in Lincolnshire near Swineshead, in the Norfolk Broads, at Monkton on the Isle of Thanet, and, of course, around Romney. Even Clerkenwell, outside London, had been built on reclaimed land. The process was necessarily a cooperative enterprise involving a multitude of people. Some projects were sponsored by large institutions, such as the archdiocese of Canterbury. Archbishops Baldwin and Thomas Becket were well-known "embankers." More often, the seemingly endless work was initiated by groups of small landowners, since the land to be reclaimed belonged to them, rather than to any lord.

The reclaimed lands, commonly called assarts, were usually

4. For a discussion of drainage in Kent, see Smith, *Canterbury Cathedral Priory*, pp. 150–51, 166–89, 203–4.

measured in acres rather than in strips, as the older open-field manorial holdings were measured. When the dimensions of post-mill sites were given, the measurements ranged from a quarter acre to more than two acres per mill and thus exceeded strip measurements. The implication seems clear: most windmills were built either on new land, or on land belonging to small landowners that was not divided into strips.

A definite relationship seems to have existed between land reclamation, small landowner enterprise, large estate management, and new windmills. The progressive thing to do was to erect a post-mill on one's holding. Theoretically, such a labor-saving device made good business sense, but whether most sponsors were motivated by such investment strategies is difficult to determine.

It was the functional similarity of the windmill to the water mill that allowed the new apparatus to glide so silently into history. The invention of the post-mill did not result in creation of an equally novel process, occupation, or product. An existing rotary device was simply powered by an alternative energy source. Thus, the windmill was never considered a distinct entity but merely another way of grinding the grain. Perhaps that was why many of the first windmills were merely called mills. Only after their popularity increased did the terminology become specific.

One windmill promoter quite deliberately used the new technology to increase his profits. Sometime before 1222, perhaps as early as 1192, Richard of Stapleford, a cleric and estate manager, built a post-mill on a manor at Ardleigh, in Hertfordshire, belonging to the canons of Saint Paul's Cathedral, London. Richard was a troubleshooter for the canons and his task was to take over unprofitable manors and whip them into shape. A survey of cathedral estates conducted in 1222 documents his success. At Ardleigh he increased the acreage under cultivation both by reclaiming wasteland and by sacrificing part of the pasture and woodland. He also erected new farm buildings, fertilized the fields, and built the windmill. The new mill constituted more than 25 percent of the renovated manor's total worth.[5] After returning Ardleigh to profitability, Richard went on to Wicham in Essex, another manor owned by Saint Paul's. There he also built a windmill and reclaimed and fertilized the land.[6] By 1222 the cathedral canons had at least three other windmills on their manors.[7]

The few economic statistics that are available on post-mills erected before 1200 are frustratingly vague. Nevertheless, if they are broadly divided into mill values and mill yields, and if all entries are converted into pence, some elementary accounting can be attempted. The total value of a windmill can only be approximated. The Ecclesdon mill was sold for 25 marks (4,000 pence) in about 1197. A bribe in the same amount was intended to assure the Christ Church monks control of a Monkton windmill. At the end of the century, the Wigston Parva post-mill was valued at 20 marks (3,200 pence). The Tostock windmill was transferred for a consideration of 720 pence. The average weekly wage of an unskilled worker in 1200 was 12 pence—the approximate cost of one sheep. This gives some idea of contemporary purchasing power.

In the late twelfth century, 96 pence was required to settle a windmill dispute between brothers in Hienhel; 160 pence, the price of a hawk, was paid to obtain possession of two post-mills in Newcastle. These figures represent the partial value, rather than the full value, of any mill. Still, one must not exaggerate a mill owner's expenses. Dean Herbert erected his windmill very quickly. He thriftily saved the lumber from its dismantling and did not complain about financial loss.

Annual profits varied. The Dunwich mill was expected to yield 80 pence a year; the Weedley mill, a high 96 pence; and the Ecclesdon mill, the equivalent of a pound of pepper. Presumably

5. *Domesday of Saint Paul's*, p. 21; Gibbs, *Early Charters*, p. 59, charter 82.
6. *Domesday of Saint Paul's*, p. 33.
7. Ibid., pp. 1, 28, 69. One mill needed repair and was probably old.

23, 24.
A woman brought grain to a miller,
felt cheated, and returned to burn down the mill.
Note the curious undercarriages of these post-
mills; perhaps they were intended to stabilize
the cabin. From Royal MS. 10 E IV
(Smithfield Priory Decretals), fols. 70v and 71.

these figures refer to profits after the deduction of expenses, including tithes. The Twigrind water mill in Hempnall yielded 160
pence.

The fees paid for the use of some windmills seem to have been
based on yearly rentals rather than full profits. A fee of 12 pence
was asked for the Brampton and Hienhel mills; the fee for the
Fincham mill was 4 pence; and the fee for the mill at Wigston
Magna was, almost nominally, a penny a year. (The last figure
may have been a ground rent only.) The Reculver and Westhallimot mills were considered profitable enough for their tithes to
be judged an appropriate hospital donation. The overall tone of
the post-mill charters seems to indicate that proprietors were as
pleased by the convenience and efficiency of their new facilities as
by their profits.

In the same manuscript, copied in Italy but
illuminated in England between 1325 and 1350,
representations of four other post-mills
illustrate tales of a giant attacking a castle
(Figure 19) and of a friar seducing
a miller's wife; see fols. 114, 115, 115v.
Reproduced by permission of the British Library.

The mechanism was readily understood and relatively inexpensive. Its basic energy source, Dean Herbert's "free benefit of the wind," was available to everyone. However, some wealthy proprietors sought to consolidate their gains by eliminating adverse competition. Some even tried to rationalize ways to evade ecclesiastical tithing. From their viewpoint, there were also disturbing, almost revolutionary, undercurrents to post-mill developments. For others, establishment of the new facilities, like the concurrent flight of serfs to towns, provided an opportunity to circumvent traditional regulations, to escape manorial prerogatives. Perhaps it was no accident that the contemporary slogan, "Town air makes one free," was so reminiscent of Dean Herbert's cry, for the average person expressed his conception of liberty in terms of vital limitless energy.

The apparent quantum leap in English windmill construction after 1180 may have been inevitable, but it was affected by the wholly unexpected rampant inflation. After decades of relative stability, the price of grain and livestock soared from about 1180 until 1220; the prices of some items doubled, or even trebled.[8] The populations of all West European nations were increasing, but only England experienced such a stratospheric rise in prices. Rent for the Tostock windmill increased by more than 30 percent in that period. This inflationary spiral, at least in food prices, was comparable only to those of the sixteenth and twentieth centuries—also unusually creative periods.

Invention of the windmill brought both advantages and disadvantages. For the mill owners, there was cash to be made by increasing operations. The new mechanical contraptions were good investments. They were easier and less costly to erect than water mills. They did not cause wasteful flooding of valuable arable land and, although dependent upon unpredictable breezes, they could continue to operate throughout the icy depth of winter. They utilized only about 10 percent of the air's available energy, but they were still remarkably efficient and infinitely perfectible. Once millers switched from sunken peg supports to exposed cross trees resting on raised piers, their maintenance costs declined.

The windmill was an instrument of social progress. It enlarged the community of skilled mechanics and lightened the daily work load of countless women. The use of windmills or water mills to grind grain into flour meant the saving of one day's labor of one person in every five. However, a disadvantage was that millers often claimed one-sixteenth of the grain brought to them as a fee for their services. Thus, to give his wife more free time, a peasant had to surrender almost 7 percent of the grain he

8. W. H. Beveridge, "The Yield and Price of Corn in the Middle Ages"; D. L. Farmer, "Some Price Fluctuations in Angevin England"; Harvey, "English Inflation"; Miller and Hatcher, *Medieval England*, p. 45; Postan, *Medieval Economy and Society*, p. 231.

counted on to feed his family. Some couples tried to compensate by wifely sewing or more intensive cultivation of the household vegetable garden. Others continued to mill at home by hand.[9] Some landlords tried to prevent this traditional practice by confiscating hand querns, but they were seldom successful.[10]

Mechanical production had another mixed side effect. Fulling mills began to appear in England about 1185. This adaptation of the water mill was already well known on the Continent but had been slow to cross the Channel (just the reverse of the windmill). The fulling mill eliminated many steps in preparing cloth, since before its invention, the cloth had had to be pounded by human feet. The use of this mill began the process of industrialization, but it also threw large numbers of unskilled workers out of work, at a time when food prices were out of control.[11]

Inflation also devastated two other groups. Landlords who leased many of their properties at fixed rents saw their profits evaporate. Even harder hit were the free peasants. The dramatic surge in prices struck them like a thunderbolt. It frustrated recent

9. This percentage was worked out by Sylvia Thrupp; see her "Medieval Industry, 1000–1500," in Cipolla, *Fontana Economic History*, 1:234–35. The introduction of the windmill probably released some animal power for other types of work; see J. C. Russell, "Population in Europe, 1000–1500," ibid., p. 68.

10. Between 1131 and 1140, Cecily de Rumilly denounced hand mills and threatened those who did not use the local mill. See Farrar and Clay, *Early Yorkshire Charters*, 7:55–66, charter 4; Lennard, *Rural England*, p. 281. The burghers of Newcastle were specifically granted the right to possess hand mills, however; see Douglas and Greenaway, *English Historical Documents*, 2:970–71.

11. On the fulling mill, see Carus-Wilson, "An Industrial Revolution of the Thirteenth Century"; Reginald Lennard, "Early English Fulling Mills: Additional Examples"; and his "An Early Fulling Mill: A Note." Each of these scholars identified three twelfth-century examples. None of the six could definitely be dated before 1185. I can add three others, also from the late twelfth century: one at Leominster, Herefordshire (B.L. Harley MS. 1708, fol. 280v, mentioned between 1189 and 1193); one at Eaton, Leicestershire (Nichols, *History and Antiquities*, vol. 2, pt. 1, appendix, p. 91); and another at Wllneford, or Welleford, for use by the monks of Bordesley Abbey; P.R.O., Ancient Deeds D.L. E 315/43, charter 39; D.L. E 315/46, charter 1. R. A. Donkin, in *The Cistercians: Studies in the Geography of Medieval England and Wales*, p. 138 and appendix V, indicates that fulling mills might have been located at the Newminster and Quarr Abbeys in the twelfth century.

attempts to expand their personal holdings and forced many into debt merely to pay annual expenses. If they failed to redeem their loans, they suffered humiliating penalties. Many proud peasants fled to the towns or found themselves degraded to manorial tenants, that is, serfs.

Inherited feudal privileges proved stronger than the progressive influence of a new technology. Some audacious proponents of the free use of wind power, such as Dean Herbert, may actually have been desperate defenders of declining liberties, rather than daring advocates of a new cause. The windmill was born in an optimistic era, but it matured at a time when increasing seignorial control was a bleak reality. It would be useful to many people in the centuries to come, but in 1200 its greatest beneficiaries seem to have been those fortunate enough to own or operate one and, of course, the ubiquitous lawyers and bureaucrats.

William the Almoner, William of Amundeville, and the cooperative villagers of Wigston Magna epitomized the generosity characteristic of the peaceful era during which the windmill was originally perfected. Late-twelfth-century investors, such as Archbishop Hubert, Abbot Samson, Doctor Gilbert, William Wade, Bishop Seffrid, Walter Fitz Robert, and Adam of Charing were tougher pragmatists. Beset by rampant inflation, increasing population, and tenant unrest, they saw technological improvement as a cost-effective way of maximizing their profits. These windmill proprietors were skillful administrators of their holdings and were ready to experiment with new techniques of estate management, new reclamation schemes, and innovative machines. The fact that others might suffer as a result of their progress was not a deterrent. However, a portion of the windmill profits usually returned to the larger community in the form of gifts to churches, monasteries, and hospitals.

Although the ties between mechanical technology and feudal society were significant and surprising, no early poet, historian, pioneering millwright, profiteering owner, brawny miller, or troubled customer wrote about them. Such insights as we do gain therefore necessarily emit the dank dust of antiquated documents rather than the fragrant odor of hard hammered oak. Yet the vibrant individuals who first promoted the windmill transcended this limitation. William the Almoner may have been an obscure cleric, but he helped broadcast an invention that challenged the foundations of medieval society and would continuously benefit all humanity.

# Appendix

## A Gazetteer
## of Twelfth-Century Windmills

This gazetteer documents the existence of fifty-six twelfth-century English windmills, far more machines than formerly known. Since the medieval terminology was so varied, the Latin phrase used to describe each post-mill is included. Some unusual spellings of place names are also specified. Few windmills reveal their construction dates. Most can only be identified as having been built before a certain terminus (e.g., −1137) or in an approximate year (c. 1185).

Although some repetition is inevitable, most of the personalities and structures described in the following pages have not been discussed in Chapters 1–9. Some of these individuals, such as the proprietor's wife who was captured by Eustace the pirate monk, stir the imagination. But what is significant is the realization that, collectively, these windmill pioneers forged a technological revolution, the first of its kind in a millennium.

### 1. BEDFORDSHIRE: RENHOLD   −1166

Some post-mills seem to spring out of the surviving records. In other cases, determining the type of mill requires laborious research. An example is the mill at Renhold, just north of Bedford. Only one milling facility was ever mentioned as existing there. In 1166 it was described merely as a mill. Before 1200 it was identified as a windmill. In 1260, it was again called simply a mill.

In 1165/66 Simon de Beauchamp, a wealthy Bedfordshire knight, replaced the secular canons of the Collegiate Church of Saint Paul in Bedford with a community of Augustinian canons. Fifteen years later these men were relocated in nearby Newnham and their establishment was henceforth called Newnham Priory. In his initial endowment, made about 1166, Simon donated to the reconstituted house, among other things, the tithes levied on the profits of seven undifferentiated mills, including one at Renhold.[1] His grant was confirmed by King Henry II between 1180 and 1189, but in that document the royal scribe also failed to distinguish between windmills and water mills.[2] As lord of the area, Simon felt free to bestow tithes from all the mills in his domain on whatever institution he pleased. Whether the secular canons who preceded the Augustinians had also received these tithes is unknown, for the priory cartulary does not record benefactions before 1166.

Between 1180 and 1200, Gilbert Avenel donated several parcels of land in Renhold to the Newnham canons. Included were the land on which the windmill stood (*terram super quam molendinum venti stat*) and half an acre next to the road which adjoins that mill.[3] The Renhold windmill was evidently not part of Gilbert's gift, for the common practice was to donate mill structures separately from the ground on which they stood. Presumably he kept the machine for himself. Little more is known about Gilbert Avenel except that he made other small grants to Newnham Priory, as well as to Croxton Abbey and Wardon Abbey, and donated a mill to Leicester Abbey.[4] He died sometime between 1198 and 1216, leaving a widow, Clarice.[5]

In the late twelfth century Geoffrey, the son of Ralph of Renhold, referred to the installation as a landmark. His charter cryptically mentioned two selions of land "toward the windmill adjoining W(alter) Rufus" (*versus molendinum venti iuxta W. Rufum*)—an odd, incomplete phrase—and four selions next to that mill. This charter was probably drawn up considerably before 1200, for early in the thirteenth century Geoffrey's grandson also used the Renhold mill as a territorial reference; he did not specify that it was a windmill.[6]

Twelfth- and thirteenth-century charters consistently refer to only one mill in Renhold.[7] Moreover, tithe records show that in 1260 the tithe, levied at a high annual rate of one-half mark, was paid only for one unspecified mill.[8] The conclusion is unavoidable: There was a solitary mill in Renhold. It was not mentioned in Domesday Book, but it was built sometime between 1086 and 1166, and it was a windmill.

By 1262, the plot beside the Renhold mill had been given the Middle English name *Wynmulnefeld*, or Windmill Field.[9] In the late thirteenth century, that name was also used to refer to a field in Salpho, not far from Renhold, which was the site of a post-mill that cannot be otherwise dated.[10] Other land in Salpho was referred to as *Wyndmulnefurlong*.[11]

In summary, in about 1166 Simon de Beauchamp granted Newnham Priory the tithes collected from the Renhold post-mill; a few years later Gilbert Avenel gave the priory the land on which the mill stood. It is not certain when the abbey acquired the windmill itself.

1. *Newnham Cartulary*, charter 5; this was confirmed by his son between 1220 and 1260 (charter 1).

2. Ibid., charter 39.

3. Ibid., charter 368.

4. Ibid., charters 364–67; *H.M.C. Rutland*, 4:177; B.L. Additional MS. 4935 (Croxton Abbey Register transcripts), fols. 96, 105v–6; *Wardon Cartulary*, charters 191–93; B.L. Cotton MS. Vitellius F XXVII (Leicester Abbey Rental), fol. 20.

5. *Newnham Cartulary*, charter 371. During King John's reign, she contested the claim of Walter Avenel (evidently Gilbert's brother) to some land. On Walter, see charters 397, 437. Gilbert appears in Curia Regis Rolls, 6 Richard I (1194-1195), p. 53.

6. Ibid., charters 386, 391.

7. For example, ibid., charters 6 (1189–1206), 13 (early thirteenth century), 1 (1220–1260).

8. Ibid., charter 111.

9. Ibid., charter 355.

10. Ibid., charter 353. It is possible, but unlikely, that the scribe erroneously wrote "Salpho" instead of "Renhold."

11. Ibid., charter 354.


## 2. BUCKINGHAMSHIRE: DINTON    c. 1184

Agnes of Mountchesney, a daughter of Payn Fitz John, one of Henry I's top officials, was a strong-willed woman who insisted on exercising personal control over her lands and possessions. She and her sister, Cecily, the countess of Hereford, were among the most learned women of the kingdom. They were proud of their father's achievements and often emphasized that they were his daughters by using their maiden names rather than their married names. Agnes was a benefactor of such religious communities as Stoke-by-Clare in Suffolk, where she held some land, and Oseney Abbey in Oxfordshire.[1] She was most supportive of the Benedictine nuns of Godstow Abbey, near Oxford.

Godstow was well known in the 1170s as the refuge of "Fair Rosamund" (d. 1176), Henry II's beautiful mistress who once enlivened nearby Woodstock Bower. Rosamund's father, Walter Clifford, gave Godstow Abbey a mill at Frampton, on the Severn, in memory of his wife and daughter. The charter contained a clause specifying that Walter also donated the land on which the mill was located. This suggests that the mill was a windmill, but later references prove that the Frampton mill was a water mill.[2]

At different times before 1182, Agnes granted to the nuns land, tithes on hay, and a parish church—all of which were located in or derived from her manorial village at Dinton in Buckinghamshire; one of her particular intentions was to support the nuns' infirmary.[3] After 1184 (probably shortly thereafter), she gave Godstow Abbey her windmill (*molendinum meum ad ventum*), which stood on a mound (*super hogam*) outside her village at Dinton, next to the royal road that ran from Oxford to Aylesbury. She also donated four acres that were part of the manor and allowed perpetual free entrance to and exit from the mill.[4]

This donation was attested by sixteen local worthies, most of whom were members of her own large household. They included Abbot Hugh of Oseney Abbey (1184–1205), her daughter Alicia, her marshal, two chaplains, two clerks, the Godstow Abbey seneschal, a janitor, a man named Walter who was identified as her painter (*meo pictore*—a rather unusual designation that probably meant artist), and her doctor, Master Christian (a relatively unusual baptismal name).[5]

Four of her gifts to Godstow would be recorded in the abbey's early-fifteenth-century Latin cartulary. Between 1467 and 1480, abstracts of her charters were recorded (in English) in a second convent register; the post-mill was then called a *windemille*.[6]

Agnes was sixty years old in 1185. She had been widowed three times, had had at least five children, and was apparently trying to avoid a fourth marriage arranged by the king.[7] She was still alive in 1186/87, but apparently died shortly thereafter.[8]

1. *Stoke-by-Clare Cartulary*, 1:104, charter 130a; *Oseney Cartulary*, 3:77–78. On Cecily's grant to her tutor, Walter of Bayeux, see Welbore St. Clair Baddeley, "Fresh Material Relating to Norman Gloucester."

2. *Godstow Cartulary*, 1:135, charters 156–60.

3. Ibid., 1:63–65, charters 50, 52–55. Several of these gifts were confirmed by Henry II about 1182 (ibid., 2:664, charter 886) and by Bishop Hugh of Lincoln about 1190 (ibid., 2: 647, charter 869).

4. P.R.O. MS. E 164.20 (Godstow Abbey Cartulary), fol. 41. Her other grants were also recorded in this manuscript.

5. *Oseney Cartulary*, 3:77–78, charter 1251 (1168–1184).

6. *English Register of Godstow*, 1:63–64, charter 51.

7. *Rotuli de Dominabus et Pueris et Puellis de XII Comitatibus (1185)*, pp. xxviii, xlii, xliii, 38, 50, 51, 58

8. *Pipe Rolls, 33 Henry II* (1186–1187), p. 209.


## 3. BUCKINGHAMSHIRE: EVERSHAW    −1200

Although now abandoned and plowed back to farmland, twelfth-century Evershaw was a small village held by minor vassals from the Beauchamp lords of Bedford. About 1150, parts of the land were given to nearby Luffield Priory, but later rented back to local knights.[1] From 1175 (perhaps from 1150) to about 1210 the military tenant was Hugh of Evershaw, the son of Ralph, the son of Richard. About 1200, or before, Hugh made several grants to Adam, the son of Turgis, who did him homage and service. Included were three acres lying near the windmill (*iacent versus molendinum de vento*). Hugh's son, Ralph, promptly confirmed his father's grants.[2] In both cases the post-mill was mentioned merely as a convenient landmark to locate adjacent property; ownership of the mill was never mentioned. Neither Hugh nor his son considered it a novelty. By 1220 the area was called Windmill Hill (*Winmulnel-hul*).[3] About that time Luffield Priory built two other post-mills on its neighboring lands.[4]

1. G. R. Elvey, ed., *Luffield Priory Charters*, charters 431–36, 438.

2. Ibid., charters 434, 459. The original charters with attached seals still exist and are preserved in Westminster Abbey, ancient charters 2489 and 2508. The modern editor of the priory records dated Hugh's charter c. 1200–1210, and his son's c. 1210–1220, but the witnesses to the two grants were identical and it is more likely they were drawn up at the same time, probably by 1200 because Hugh was rather elderly, having matured before 1150; *Luffield Priory Charters*, 2:151, citing an attestation to B.L. Harley Charter 85 C 21. The editor also believed the Evershaw windmill could easily be a twelfth-century creation; ibid., 2:liv. Adam of Turgis mentioned the two charters about 1225/30; charter 451.

3. Ibid., charters 447, 457.

4. Ibid., charters 73, 117.

## 4. CAMBRIDGESHIRE (?): N——   —1191 × 1196

Archdeacon Burchard of Durham once tried to collect tithes from a windmill (*de molendino ad ventum*) located in a parish where he was rector. The parish was identified only as N——. Although the mill's owner, a knight, had previously paid tithes on the land, he refused to render them on the profits of the new mill. This dispute may have continued for some time, because two knights were mentioned in connection with the windmill: Knight M—— constructed it, but Knight H—— was eventually ordered to pay the tithe.

Archdeacon Burchard appealed to Rome, whereupon Pope Celestine III (1191–1198) directed the archdeacon of Ely and the abbot of Ramsey to ensure that the tithes were paid. Celestine's command was regarded as a precedent for all such disputes and thus became a standard entry in canonical decretal collections.[1]

Burchard seems to have died in 1196.

1. For a fuller discussion of the dispute, see Chapter 8. Mary G. Cheney, in "The Decretal of Pope Celestine III on Tithes of Windmills, JL 17620," shows that the rector was Archdeacon Burchard, even though the texts merely refer to him as B——.

## 5. CAMBRIDGESHIRE: SILVERLEY   —1166 × 1194

Between 1166 and 1194, probably about 1166, Reginald Arsic donated to the Priory of Hatfield Regis in Essex the tithes from his windmill (*molendini mei de vento*), located in the field of Breche in Silverley. Reginald's original deed (with his seal) for this grant still exists.[1] His charter is probably the oldest original document to include mention of a windmill. His tithes were amicably paid, whereas those owed to Archdeacon Burchard were refused.

1. B.L. Additional Charter 2834. See Figure 15. This transaction is discussed in Chapter 4. The Templar survey, which recorded the Weedley windmill in Yorkshire, was conducted in 1185 and transcribed shortly thereafter. Original charters pertaining to donations of windmills at Evershaw, in Buckinghamshire, Little Foxmold, in Kent, and Manby, in Lincolnshire, were also written before 1200. Archdeacon Burchard's tithes are discussed in Appendix entry no. 4.

## 6. ESSEX: HENHAM   —1200

In 1202/3, before the royal court at Chelmsford, William Fitz John quitclaimed a number of parcels of land in Henham, including one and one-half rods at the windmill (*molendinum de vento*). Thomas Ivelchild was also involved in the case.[1]

William was a valued retainer of Walter Fitz Robert, lord of Dunmow and patron of several post-mills. Before 1187 William had attested Walter's settlement with regard to the Drayton windmill.[2] Henham was part of the marriage portion of Matilda d'Oilli, Baron Walter's second wife. As his widow, she reconfirmed her rights in court between 1199 and 1200.[3] Tithes from all the Henham mills had been given to Dunmow Priory early in the twelfth century, so the windmill probably existed long before 1200.[4]

1. *Feet of Fines, Essex,* 1:29, item 119.
2. *Christ Church Cartulary,* p. 54; also see Chapter 5.
3. *Feet of Fines, Essex,* 1:19–20, item no. 13.
4. B.L. Harley MS. 662 (Dunmow Priory Cartulary), fols. 6–7, 57v. Bishop Gilbert Foliot confirmed some of these tithes between 1167 and 1180; Adrian Morey and Christopher Brooke, *The Letters and Charters of Gilbert Foliot,* charter 368.

## 7–8. KENT: CANTERBURY   c. 1200

Milling facilities were located in Canterbury and throughout Kent. Seven fine, long-established water mills functioned on the dual branches of the River Stour—five in the center of town and two beyond the town's walls. There was a horse mill at Westgate. These would be joined by numerous suburban post-mills, at least two of which were built in the twelfth century. Several others may also have been constructed then, but they are difficult to date.

The Canterbury windmills were all owned by religious houses— the Abbey of Saint Augustine, the affiliated Saint Lawrence Hospital (a leprosarium), Saint Sepulchre's Convent, and the Hospital of Saint Thomas the Martyr at Eastbridge. Energetic leaders of these institutions, especially Abbot Roger of Saint Augustine's Abbey (1175– 1212), were interested in the new power technology. Christ Church Cathedral Priory, from which Roger had come, probably also had local windmills, but such engines have only been located on its manors in the Kentish countryside.

The two earliest datable Canterbury windmills stood north and east of the city—in Northwood, a forested area to the north, and beyond Ridingate, on the south side of Old Dover Road near Saint Sepulchre's Convent, and the leprosarium of Saint Lawrence, on the first eastern rise outside town. The last windmill in the Ridingate suburb burned down in 1873. One wind-powered installation still stands northeast of the city, on Saint Martin's Hill, but whether it has early medieval predecessors is unknown.

The most interesting example is an early cooperative venture between the Saint Sepulchre nuns and the Eastbridge Hospital brethren. The convent had been founded by Archbishop Anselm about 1100, but it had only modest endowments, including some land from Saint Augustine's. Eastbridge Hospital had a considerable interest in several windmills and water mills, such as the tithes from post-mills at Reculver and Thanet (Appendix entries nos. 11 and 13).

About 1200 the nuns entered into a contract with the hospital master and brethren. The prioress's copy of the detailed agreement still exists but, oddly, does not identify her by name. It granted Eastbridge Hospital one-quarter acre in the western part of Little Foxmold, to be used for construction of a windmill (*ad construendum molendinum suum ad ventum*). The hospital was to pay the nuns six pence a year, a high rental for so small a plot. The nuns would furnish 25 percent of the cost of building, maintaining, and repairing the mill. In turn, they were to receive the same percentage of the profits and to have their grain ground there whenever they pleased. The miller was instructed to act on behalf of both the convent and the hospital, and the latter would make a path leading from the main road to the mill.[1]

A second charter, prepared by Prioress Juliana of Saint Sepulchre's Convent (who probably composed the previous agreement as well), recorded the grant of one-quarter acre in Little Foxmold to the rector and brothers of Eastbridge Hospital. The only version of the deed that now exists is a printed excerpt that does not mention a windmill but includes the names of some witnesses; the deed can be dated between 1195 and 1206.[2]

About the same time or shortly thereafter, Richard of Bramford, the steward of Christ Church, gave the Eastbridge Hospital "whatever he had in the windmill at Foxmold" (*quicquid habui in molendino ad ventum apud Foxmolde*).[3] Foxmold and Little Foxmold were adjoining plots on the south side of Old Dover Road, divided by Ethelbert Road.

Richard of Bramford considered the Foxmold machine an investment. He did not name his associates. His career probably included a legal practice as well as estate management, for in 1206 a Richard of

Bramford, an attorney representing the bishop of Evreux, presented a plea before the king's court in Suffolk.[4] In the 1220s he leased property from Saint Augustine's Abbey, including a mill (unspecified) in Sturrey.[5]

Osmund Pollre, who attested Richard's charter and both charters of the Saint Sepulchre nuns, owned three acres in Foxmold. He described them as next to the mill of Saint Thomas and that of Richard the Seneschal, and adjacent to the lands of the nuns to the west. He did not specify windmill, although obviously that was intended.[6] Lambin Fleming, a witness to the nuns' first grant, was a merchant who had a stone house next to Eastbridge Hospital and established a fund for the illumination of that institution's infirmary. He was a friend of Osmund Pollre and was active from at least 1163 to 1240–1247, unless this span included part of the life of an unidentified son of the same name. Lambin owned a variety of city properties as well as an unspecified mill in Taynton.[7]

Most of the land in the suburb east of Canterbury belonged to the monks of Saint Augustine's Abbey. In 1137 Abbot Hugh established Saint Lawrence Hospital there to accommodate a chaplain and sixteen men and women patients. The abbot's grant consisted of nine acres, as well as the tithes levied on the abbey's large manor, called Longport, and on its land on the right side of the road from Canterbury to Dover.

Sometime thereafter, probably before 1208, a post-mill was raised for the benefit of the lepers. It was mentioned only as a landmark when the hospital acquired from Alfred Lymberner, son of Norman, the land extending from the mill to the land of Terry the goldsmith, a well-known Canterbury minter.[8] The structure was not specifically called a windmill, but its upland location suggests that it was one. Other investigators seem to have agreed.[9]

There may have been one more post-mill in this area. In about 1200 an unspecified mill was recorded at Burgate, far from any watercourse. Burgate is the next gate north of Ridingate, and this mill was evidently within the town walls. If it was a windmill, it must have been a small one, perhaps atop a building.[10] All these mill sites on the eastern side of town were close to one another.

Canterbury's water mills suffered considerable damage during the anarchy of King Stephen's reign. Shortly after he became king, therefore, Henry II ordered a panel of twenty-four men to report on their condition. He wanted to ensure that no mills, either inside or outside the city, were raised so high as to damage the monks of Christ Church. The meaning of his command is obscure. It most likely refers to water levels, but it might refer to local wind-powered mills.[11]

Sometime between 1199 and 1216, the monks of Saint Augustine acquired a post-mill (*molendinum ad ventum*) and its half-acre site, just north of the city, from Stephen Fitz Jordan of Scapeya. This donor wished to establish a private chapel to ensure that prayers would be offered for his parents, his wives, his brother, and the two kings who had given him land for his service, Richard and John. He therefore granted to the abbey two acres in Northwood for construction of the chapel and ten acres to feed and clothe its chaplain. He also granted the lesser tithes from his tenements, a windmill, and the half-acre plot on which the mill was located. These twelve and one-half acres and the mill were expected to yield seven pence at both Easter and Michaelmas.[12] The implication seems to be that the monks received the land and the post-mill, but someone else was to manage and rent them.

In 1190 Abbot Roger had given Stephen some rights in Northwood, perhaps upon the order of King Richard. Now his favor was partially returned.[13] Roger may have deliberately engineered this whole transaction. He was the main force behind the construction of one of the abbey's other post-mills, at Chislet near Reculver.[14]

Another windmill belonging to the monks of Saint Augustine was located on their manor, Elmstead, about eight miles south of Canterbury. The post-mill (*molendinum ad ventum*), its site, and the road to it were granted by John of Wadesole during the tenure of Abbot Roger. However, there were two Johns of Wadesole (father and son), and two abbots named Roger—Roger I (1175–1213) and Roger II (1253–1273). The latter Roger seems more likely; thus the windmill cannot be dated before 1253.[15]

1. Canterbury Cathedral Archives, Eastbridge Hospital Muniments, Ancient Charter C 3 (a chirograph). The seal is missing, but seven local witnesses are named. One, Lambin Fleming, was mentioned in various documents from 1163 to 1247. Another signatory, Osmund Pollre, was mentioned from 1206 to 1230. Robert the chaplain who witnessed this document is most likely Robert the clerk of Saint Sepulchre who attested for Prioress Juliana between 1180 and 1190; Canterbury Cathedral Muniments, Ancient Charter C 1161; printed in William J. Urry, *Canterbury under the Angevin Kings*, p. 419. In 1640 William Somner paraphrased the Little Foxmold windmill charter in *Antiquities of Canterbury*, p. 119; he cited a ledger of Eastbridge Hospital

that I have been unable to find. Somner added a phrase stating that the windmill was in the hundred of Ridingate. In 1775–1776 Joseph Strutt, in *Honda Angel-ynnan, or A Complete View of the Manners, Customs, Arms, Habits, etc. of the Inhabitants of England,* 2:13, quoted Somner's summary and added that it was the first detailed description of a windmill that he knew of. In 1785 John Duncombe and Nicholas Battely published most of the windmill charter and corrected Somner's epitome in *The History and Antiquities of the Three Episcopal Hospitals at or near Canterbury: viz., St. Nicholas, at Harbledown; St. Johns', Northgate; and St. Thomas, of Eastbridge,* p. 312. In 1798 Cyprian Bunce completed his handwritten catalog of the Eastbridge Hospital muniments and included the charter; A Register of Loans and Charitable Donations to the Poor of Canterbury, 2:694–95. In 1899, Richard Bennett and John Elton, in *History of Cornmilling,* 2:235, mentioned Strutt's final remark and gave a confusing citation to his work. This is merely a brief outline of the writ's passage, for the works of Somner, Duncombe and Battely (himself a master of the Eastbridge Hospital), and Strutt were published in various editions.

2. Duncombe and Battely, *Three Episcopal Hospitals,* p. 313. I could not find this charter in any records maintained by Eastbridge Hospital or Saint Sepulchre's Convent. The Eastbridge Hospital muniments have lost a few charters over the years and this must be one of them. B.L. Additional MS. 5516 (Holy Sepulchre Priory Cartulary) does not include mention of any windmill. There were two prioresses named Juliana, one known in about 1184 and another known from about 1227 to 1258; David Knowles, C. N. L. Brooke, and Vera C. M. London, *The Heads of Religious Houses: England and Wales, 940–1216,* p. 210; Urry, *Canterbury,* pp. 418–19. The second charter was attested by Archdeacon Henry of Canterbury. There were also two Archdeacon Henrys: Henry de Castilion, c. 1195–1206, and Master Henry de Sandford, 1208–1226, when he was elected bishop of Rochester. The second cleric usually attested as Master Henry, so the first probably attested this charter. The four other witnesses, including Osmund Pollre, also attested the unnamed prioress's chirograph.

3. Canterbury Cathedral Archives, Eastbridge Hospital Muniments, Ancient Charter C 4. Part of a seal is still attached to the charter. Calendared in Bunce's catalog, 2:695; printed incompletely in Duncombe and Battely, *Three Episcopal Hospitals,* p. 312. One witness to the grant, William Crevequer, resigned as parson of Blean Church before 1228, perhaps before 1206; Somner, *Antiquities of Canterbury,* p. 61; S. Gordon Wilson, *With the Pilgrims to Canterbury and the History of the Hospital of Saint Thomas,* p. 40; Kathleen Major, ed., *Acta Stephani Langton, Cantuariensis Archiepiscopi, A.D. 1207–1229,* p. 155. Since Richard of Bramford's windmill at Foxmold cannot be definitely dated before 1200, it is not included as a separate entry in this Appendix.

4. Doris M. Stenton, ed., *Pleas before the King and His Justices, 1198–1212,* 3:270, item no. 1960.

5. On Richard, see Canterbury Cathedral Muniments, Ancient Charter C 1088 (a writ attested for Robert the janitor); *Register of Saint Augustine's Abbey, Commonly Called the Black Book,* pt. 1, pp. 131, 134 (the Sturrey mill), 139, 166, 168, 170 (an attester in 1221), 389, 420, 485 (a witness to a deed concerning the grant of a mill in Lenham). The name Richard of Bramford is similar to that of Master Richard of Brandeston, who may have been the chancellor of Archbishop Hubert Walter (1193–1205). On this Richard, see Christopher R. Cheney, *Hubert Walter,* p. 164. He attested the archbishop's grant to the Eastbridge Hospital of tithes from windmills at Reculver and Thanet. A

knight of the Temple, Brother Richard of Bramford, was master of Sutton preceptory in 1251; *Kentish Cartulary*, p. 106.

6. Canterbury Cathedral Muniments, Register A, Fol. 433v. He died before 1247. When his father, Robert Pollre, held the property, it was described as near Saint Lawrence Hospital (fol. 434v). For other references to Osmund in the years 1206–1230, see Canterbury Cathedral Muniments, Ancient Charters C 712, C 1088; *Saint Gregory's Cartulary*, charters 81, 85, and 225 (where he was described as a prefect of Canterbury); *Register of Saint Augustine's Abbey*, pp. 341, 391 (again described as prefect); Urry, *Canterbury*, pp. 332 (c. 1206, where Osmund is said to have held four acres in Foxmold), 415 (–1227).

7. On Lambin, see Canterbury Cathedral Archives, Eastbridge Hospital Muniments, Ancient Charters A 15 (his charter), A 70, A 71, F 20 (all attestations before 1231); *Saint Gregory's Cartulary*, pp. 147–48, 163, 170 (in 1240), 205; Urry, *Canterbury*, charters 37 (–1227), 70 (1220); rentals, pp. 236 (c. 1163–1167), 259 (1200), 239 (1206), 343, 348, 358–59, 375, 378; *Register of Saint Augustine's Abbey*, pt. 1, pp. 336, 342; pt. 2, 581, 592; *Feet of Fines, Kent*, p. 196 (a 1247 complaint of his widow, Margery, and his son, Thomas, about the Taynton mill). The records do not suggest that more than one Lambin was involved in any of these transactions.

8. Canterbury Cathedral Muniments, Literary MS. C 20 (Saint Lawrence Hospital Register), fol. 57; Oxford University, Bodleian Library MS. Topham Kent, d.3 (Saint Lawrence Hospital Register), fol. 24v. The charter was printed in C. Eveleigh Woodruff, ed., "The Register and Chartulary of the Hospital of St. Lawrence, Canterbury," p. 47. Terry died before 1214, probably before 1208; Urry, *Canterbury*, pp. 12, 174–76. Alfred Lymberner was also mentioned in about 1220; *Register of Saint Augustine's Abbey*, pt. 1, p. 389; and in an undated context—Woodruff, "Register", p. 47; and Canterbury Cathedral Muniments, Literary MS. C 20, fol. 86. He died before 1238–1242; *Saint Gregory's Cartulary*, p. 152, charter 211. One witness to his mill-defined gift was William Conkin. A philanthropist of that name founded a Canterbury hospital about 1190, but someone else of the same name was still active in the 1240s; Urry, *Canterbury*, pp. 52, 58, 120, 191, 211, 225.

9. Woodruff ("Register," p. 50) clearly thought the structure was a windmill. Thomas de Bery also vaguely mentioned the mill in 1270. He granted to the hospital land by the road that leads to the mill; Canterbury Cathedral Muniments, Literary MS. C 20, fol. 50; Oxford University, Bodleian Library MS. Topham Kent d.3, fol. 22; calendared in Woodruff, "Register," p. 46. Because this post-mill is difficult to date precisely, it has been omitted from the Appendix.

10. Urry, *Canterbury*, p. 263.

11. For information on Canterbury mills under Stephen, see *Regesta*, vol. 3, charters 163–65. On Henry II's order, see Canterbury Cathedral Muniments, Ancient Charter C 23; printed in Elie Berger, ed., *Recueil des actes de Henri II concernant de provinces françaises et les affaires de France*, 1:208.

12. *Register of Saint Augustine's Abbey*, pt. 2, pp. 503–4.

13. Ibid., p. 502.

14. See Appendix entry no. 9.

15. B.L. Cotton MS. Claudius D X (Red Book of Saint Augustine's Abbey), fols. 215–15v. One witness to the grant, Master Hamo Doge, was prominent in the mid-thirteenth century and received several grants from Abbot Roger II between 1266 and

1268; *Register of Saint Augustine's Abbey*, pt. 1, p. 228; pt. 2, pp. 563–67. John Wadesloe the younger often identified himself as John, the son of John. This donor simply called himself John of Wadesloe.

## 9. KENT: CHISLET   c. 1200

Abbot Roger of Saint Augustine's Abbey, Canterbury (1175–1213), was an energetic administrator. He built a windmill (*plantavit unum molendinum ad ventum*) on his monastery's manor of Chislet, in northern Kent, not far from Reculver and Thanet. It is impossible to specify when during his long abbacy this post-mill was erected, but it certainly existed by 1209.

In the tenth year of King John's reign (that is, May 1208–May 1209), a free tenant at Chislet named Peter, the son of Richard of Wischebeck, gave to the abbey about one virgate of land, which evidently constituted most of Boytun Hill. He noted that Abbot Roger had built one windmill there. Peter further agreed to provide a path leading from the public road to the mill suitable for both men and horses.[1]

1. *Register of Saint Augustine's Abbey*, pt. 2, p. 411. Although the Chislet installation cannot be positively dated before 1200, it is included in this Appendix because it was obviously part of the rapid increase in windmill construction that occurred in eastern Kent in the late twelfth century.

## 10. KENT: MONKTON, THANET   –1198

William of Wade (or Wood, or Acol), who was from near Monkton on the Isle of Thanet, built a windmill (*molendinum ad ventum*) on land that apparently owed some obligation to the monks of Christ Church, Canterbury. The prior contested its construction, and sometime during the reign of King Richard (1189–1199) obtained a court decision ordering William to demolish the post-mill.[1]

However, in 1199 William delayed, Richard was killed, and John became king. The knight then traveled to Poitou in an attempt to have the case reopened. He obtained three postponements of court action, two in Michaelmas 1199 and one in early 1200. On December 3, 1199, John confirmed construction (*possint construere molendinum venti*).[2] In 1200 William was recorded as offering the king ten marks for a royal license (after the fact) to erect a windmill (*ut possit construere molendinum venteritium*).[3] This seems to be the earliest usage of the more elaborate term for a post-mill (*molendinum ventritium* or *molendinum ventricium*). At the same time, the Canterbury monks offered ten marks to have the case with regard to William's windmill (*de molendino ad ventum*) given a new hearing.[4]

In 1202/3 William Wade appeared in the priory's court at Canterbury to attempt to settle the differences between himself and the monks. He promised in a charter not to build another mill in Monkton, or anywhere else on the Isle of Thanet.[5] Not satisfied with this declaration, and mindful of their earlier experience with William, in October 1203 the monks obtained a date for the royal hearing of their case (*de placito molendini*).[6]

Early in 1204, after paying twenty-five marks for a speedy judgment, the monks presented their formal complaint, alleging that William had obtained King John's permission to build the windmill (*molendino ad ventum*) in violation of the agreement concluded during Richard's reign, that William had not obtained their consent, and that he had deceived King John concerning the earlier agreement. The monarch ruled that the mill was to be destroyed and that no other should be erected.[7]

The mill may not have been destroyed, and Christ Church may have preserved it after all. At about this time, the monks arranged with a Gilbert Scot of Thanet for the use of an unspecified Monkton mill.[8] Perhaps they reasoned that they needed to recoup their trial expenses. William Wade's reputation was evidently not affected by the controversy, for he was subsequently sought out to participate in legal inquests.[9]

1. For an extended discussion of this post-mill, see Chapter 7. These details were revealed in the priory's formal complaint in 1204; *Curia Regis Rolls*, 3 (1203–1205):86–87.

2. *Rotuli Chartarum*, 1:36b, cited in Leopold Delisle, "On the Origin of Windmills in Normandy and England," p. 404. Doris M. Stenton, *Pleas before the King*, vol. 1, items nos. 2133 (Michaelmas 1199), 2548 (Michaelmas 1199), 2782 (Hilary 1200). Stenton admits that the dates assigned to the essoins are best guesses, so the chronology may, in fact, have been somewhat different. In truth, the logical narrative of these events would suggest that these postponements came after the monks initiated their second suit, that is, after 1200.

3. *Pipe Rolls*, 2 *John* (1200–1201), p. 215.

4. Ibid.

5. Canterbury Cathedral Muniments, Ancient Charter M 131. This deed was prom-

inently copied in Canterbury Cathedral Muniments, Register C, fol. 130; Register E, fol. 203. The charter gives the date of the agreement as the feast of Saint Gregory in the fourth year of the reign of King John (April 1202 to March 1203). There were two major saints named Gregory, commemorated on March 12 and on November 17. Therefore, the chronology cannot be definite.

6. *Curia Regis Rolls*, 3 (1203–1205):33.

7. Ibid., pp. 86–87.

8. Canterbury Cathedral Muniments, Register C, fol. 130.

9. *Curia Regis Rolls*, 2 (1201–1203):127 (in 1202), 164, 269 (in 1203), 3 (1203–1205):252 (in 1205). William was also listed in an abbey rental of 1207 as owing the monks more than five marks; Patricia M. Barnes and W. Raymond Powell, eds., *Interdict Documents*, p. 71. This case has been mentioned or discussed, but on the basis of limited evidence; see Delisle, "Origin of Windmills," p. 404; *Memoranda Roll, 1 John (1199–1200)*, p. xli; Stenton, *Pleas before the King*, 1:259; C. T. Flower, *Introduction to the Curia Regis Rolls, 1199–1230*, pp. 283–84.

## 11. KENT: RECULVER   –1195

Reculver was an ancient manor of the archiepiscopal see of Canterbury. Between 1193 and 1195, Archbishop Hubert Walter decided to increase the endowment of Eastbridge Hospital. Because he was moved "by the poverty and need in which our sons the brethren of the hospital of Saint Thomas of Eastbridge are forced to labor," he gave it the tithes on the profits of the Westgate mill in Canterbury, the mill and two salt pits in Herewick, the windmill in Reculver (*decimam molendini ad ventum in Raculfre*), and a windmill on the Isle of Thanet. The grant was attested by many members of the primate's entourage.[1]

There is no way of knowing whether the Reculver windmill had been erected at Hubert's command or at the instruction of one of his predecessors. Hubert was, however, an enthusiastic builder on all his properties. He was also a close relative of Theobald of Valeins, the active post-mill promoter in Hickling.

1. This document was originally Ancient Charter A 1 in the Eastbridge Hospital Muniments; printed by Duncombe and Battely in *Three Episcopal Hospitals*, p. 303. It was reported missing by Cyprian Bunce when he prepared his catalog of the muniments in 1798. The charter has been newly edited by Christopher Cheney, *English Episcopal Acta*, vol. 3, document 403, and dated by him as sometime between Nov. 1193 and April 1195. I am grateful to Professor Cheney for prepublication notice of this citation. Witnesses to the grant include Master Richard the chancellor, Ranulf the treasurer of Salisbury, Geoffrey de Bocland, Master Simon of Southwell, and Master Reiner. On these men, see Cheney, *Hubert Walter*, pp. 158–71; Charles R. Young, *Hubert Walter, Lord of Canterbury and Lord of England*, pp. 56–64. Archbishop Hubert's grant was subsequently confirmed by a J[ohn], prior of Christ Church; Canterbury Cathedral

Archives, Eastbridge Hospital Muniments, Ancient Charter A 2; Duncombe and Battely, *Three Episcopal Hospitals*, p. 303. This could be either Master John of Sitting-bourne (1222–1236) or John of Chetham (1237–1238).

## 12. KENT: ROMNEY   –1190

The leper hospital of Saint Stephen and Saint Thomas at New Romney was founded before 1185–1190 by Adam of Charing. It is located near the place where he prevented Thomas Becket's flight into exile. An attorney and well-known Kentish landowner, Adam made numerous gifts to his hospital, including one-half of a windmill (*molendinum ad ventum*) in the nearby marsh. Although two of his charters for the hospital have survived and a third was recopied as part of a confirmation of Archbishop Baldwin (1185–1190), the only record of his windmill gift is a grant of protection issued by Pope Innocent III in 1211, about five years after Adam's death.[1]

About 1200, Stephen Fitz Hamo gave the Romney lepers half of his mill. The other half belonged to William Fitz Christina, an associate of Adam. This second mill was not identified, but it may well have been a post-mill.[2]

Other post-mills may have existed at Charing in the early thirteenth century, as well as at Lenham, near Maidstone, in areas where Adam leased property from Saint Augustine's Abbey.[3]

1. The Romney Hospital was demolished in 1481 and its property was given to Magdalene College, Oxford, which preserved many of its charters. On Innocent III's confirmation, see Magdalene College Muniments, Ancient Charters, Romney Marsh Deed no. 31; printed in Innocent III, *Letters*, pp. 147, 263. On Adam's charters, see Romney Marsh Deeds nos. 61 (before 1185–1190), 59 (Baldwin's copy, 1185–1190), and 60 (after 1191). For the hospital's history, see A. F. Butcher, "The Hospital of St. Stephen and St. Thomas, New Romney"; and S. E. Rigold, "Two Kentish Hospitals Re-examined: St. Mary Ospring and SS. Stephen and Thomas, New Romney." On Adam's career, see Chapter 7. See also Urry, *Canterbury*, pp. 180–81; and Urry, "Two Notes on Guernes de Ponte-Sainte-Maxence: *Vie de Saint Thomas*."

2. Oxford University, Magdalene College Muniments, Ancient Charters, Romney Marsh Deed no. 37.

3. On Adam's connection with Lenham, see *Register of Saint Augustine's Abbey*, pt. 2, p. 483. For information on several windmills near Charing, see *About Charing: Town and Parish*, pp. 17, 38–40. William Thorne, the fourteenth-century Canterbury chronicler, quotes Pope Celestine as consenting to the appropriation of Lenham Church by Saint Augustine's Abbey to support its refectory. According to Thorne, immediately thereafter an unnamed abbot persuaded Sir John Mason, the vicar of Lenham, not to claim the tithes on the profits of the manor's three mills, including its windmills;

William Thorne, *Chronicle*, pp. 528–31. The pope would seem to be Celestine III (1191–1198), rather than Celestine IV (1241) or Celestine V (1294). Evidently he did not mention the windmill but rather left the arrangements concerning Lenham's appropriation to an abbot. The interval between the pope's order and the abbot's action cannot be specified.

## 13. KENT: WESTHALLIMOT, THANET   –1195

The archbishops of Canterbury had a large manor on the Isle of Thanet. Between 1193 and 1195, Hubert Walter augmented the endowment of Eastbridge Hospital, in Canterbury, by granting to the hospital the tithes from the windmill at Westhallimot (*decimam molendini ad ventum in Westhallimot in Tenet*) and from the windmill at Reculver, just across the inlet.[1] The archbishop's charter included none of the elegant terminology that was used to describe the post-mill in Monkton. Neither mill paid tithes before.

1. Duncombe and Battley, *Three Episcopal Hospitals*, p. 303; Cheney, *English Episcopal Acta*, vol. 3, document 403 (forthcoming). On the dating of the charter, see Appendix entry no. 11.

## 14. LEICESTERSHIRE: WIGSTON MAGNA   –1169

Before 1169 Robert, Earl of Leicester, confirmed to Simon Wykingeston, one mill and one acre in upper Wigston.[1] This was the only section of the village about which early records are available, and it is totally unsuited for a water mill. In 1296 only one mill, a windmill, existed there. Thus the post-mill existed before 1169—probably much earlier, since in the charter the earl mentioned two previous holders of the mill.

1. Alexander H. Thompson, ed., *A Calendar of Charters and Other Documents Belonging to the Hospital of William Wyggeston at Leicester*, pt. 2, p. 451, charter 867. For discussion of this post-mill see Chapter 3.

## 15. LEICESTERSHIRE: WIGSTON PARVA   –1137

In 1137 King Stephen confirmed to Reading Abbey the donation of the small manor of Wigston Parva and its mill, both of which had belonged to his almoner, William.[1] There was no mill in Wigston Parva in 1086. After 1137 the mill was leased from Reading Abbey successively by Arnold Wood, Peter of Belgrave, and Ralph of Arraby, whose charter explicitly identified it as a windmill (*molendinum ad ventum*). This is the earliest known post-mill.

1. B.L. Egerton MS. 3031 (Reading Abbey Cartulary), fols. 18v, 44; Harley MS. 1708 (Reading Abbey Cartulary), fols. 30–30v, 121v–122; Cotton MS. Vespasian E XXV (Reading Abbey Cartulary), fols. 24, 66v. For an extended discussion, see Chapter 3.

## 16. LINCOLNSHIRE: FRISKNEY   −1189 × 1194

Gilbert of Benniworth was a major benefactor of the Gilbertine canonesses of Ormesby (sometimes called North Ormesby), in Lincolnshire, a house founded between 1148 and 1154. Among his many gifts were two windmills, one on the coast at Friskney, and one further north, some distance inland, at North Elkington. Gilbert was cut from the same cloth as several other post-mill donors. He came from a large family, associated with physicians (including Haldane), was generous to several charities, composed a remarkable number of charters, held property in several shires (including Norfolk), and promoted the new technology.

Before 1189–1194, Gilbert gave to the Ormesby nuns in pure alms (that is, not subject to any financial or military obligations) several parcels of land (with tenants) in Friskney, as well as a saltpan and a windmill (*molendinum meum de vento*) with its site, appurtenances, and the land on either side of the mill, be it sand or high ground. Twenty-one men attested his donation, including his feudal lord, Philip I of Kyme (d. 1189–1194), and Stephen of Wilberforce, an athlete (*athleta*).[1] Gilbert's charter clearly indicates that he gave both the land and the windmill. He must have been elderly by 1189, for he had been attesting charters since at least the 1150s. This writ could therefore have been dictated long before 1189. There was a windmill in Ormesby, but it cannot be positively dated before 1200.[2] Gilbert also leased land from Ranulf, earl of Chester (1187–1232), another windmill promoter.[3]

Gilbert was the son of Sibyl and Roger of Benniworth and had a son named William. Gilbert's nephew, Walter (the son of his brother, Matthew), mentioned the site of a mill in Langton, Lincolnshire,[4] which was later the site of a windmill (*sedem molendini mei ventosi*) owned by John of Holm, the son of James of Flanders. John donated the land to Kirkstead Abbey sometime before 1231.[5]

1. Frank M. Stenton, ed., *Transcripts of Charters Relating to the Gilbertine Houses of Sixle, Ormesby, Catley, Bullington, and Alvingham*, p. 45. The editor dates the charter to the early thirteenth century, but it is clearly an earlier document.

2. *Guisboro Cartulary*, 2:268, charter 572; see also pp. 277, 419. The mill was listed in a thirteenth-century rental survey as the gift of Walter de Percy, son of William de Percy.

3. On Gilbert and Ranulf, see P.R.O., Ancient Charters, Duchy of Lancaster D.L. 25/3073, D.L. 25/3090. For information on Ranulf's windmill in Barsby, Leicestershire, see Oxford University, Bodleian Library Laud MS. Misc. 625 (Leicester Abbey Rental), fol. 20. The mill cannot be dated precisely.

4. Frank M. Stenton, ed., *Documents Illustrative of the Social and Economic History of the Danelaw*, esp. charters 221–22, 354. On Gilbert's other charters, see Appendix entry no. 19.

5. B.L. Cotton MS. Vespasian E XVIII (Kirkstead Abbey Cartulary), fol. 166; also fols. 163v, 164, 165. See also Frances Hill, *Medieval Lincoln*, p. 386. The charter between John and the abbey was recopied in B.L. Lansdowne MS. 207 E (Kirkstead Abbey Cartulary transcript), fol. 151. For more on him, see B.L. Cotton MS. Vespasian E XX (Bardney Abbey Cartulary), fols. 212v, 246–46v. Gilbert quitclaimed to Bardney Abbey all rights to the land bordering Langton (fol. 90).

## 17. LINCOLNSHIRE: HOGSTHORPE   −1200

Beatrice, the younger daughter of Ralph of Mumby, donated to Barlings Abbey in Lincolnshire, a Premonstratensian house founded in 1154, her windmill (*molendinum meum ad ventum*) and the two selions on which it was situated. The land was located in a place "vulgarly called Lichelomacre" at the entrance to Hogsthorpe, a village near the coast, about twelve miles north of Friskney.[1]

Beatrice's grant was confirmed explicitly by her older sister, Alice, and generally by her husband, Robert de Trus.[2] Evidently Beatrice had previously received the land from Walter Basilmann. The lady's donation cannot be dated precisely, but it must have been made considerably before 1200 because her nephew, Robert (Alice's son) wrote to Abbot Robert, who ruled Barlings from about 1203 to about 1222.[3]

Beatrice's father, Ralph of Mumby, may be Ralf of Munti, the donor of the Willingham post-mill in Suffolk. A Ralph of Mumby was mentioned in a Berkshire plea in 1200, but he may be someone else.[4]

1. B.L. Cotton MS. Faustina B I (Barlings Abbey Cartulary), fol. 36.

2. Ibid., fols. 39–39v. The charter referred only to a mill.

3. Ibid., fol. 36v. Frank M. Stenton, in *The Free Peasantry of the Northern Danelaw*, p. 34, summarized a charter (Barlings Abbey Cartulary, fol. 38v) issued by Peter Fitz

Gamel to Beatrice of Mumby and estimated its date to be 1220, but that seems too late.

4. Doris M. Stenton, *Pleas before the King*, 1:268, item no. 2864.

## 18. LINCOLNSHIRE: MANBY   −1200

About 1200 or before, Abbot Richard of Grimsby gave Walter, the son of Alan de Burc, several parcels of land including a toft in Manby with a windmill (*molendinum ad ventum*).[1] This is a rare example of a religious institution granting a post-mill to a lay person. A second, seemingly earlier charter, written by the same scribe and attested by the same eleven witnesses as the first, is a warranty regarding the same lands from Elias, the son of John of Karleton, to Walter Fitz Alan de Burc.[2]

Both charters seem to suggest that the property, and presumably the windmill, had previously been given to the Augustinian canons of Grimsby (or Wellow Abbey as it was sometimes called) by Gilbert Fitz Roger. The second charter referred to the windmill, its site, and sixteen acres, for which Walter was to pay the canons a rent of twelve pence. The first charter referred to the windmill, its site, and nineteen acres, for which Walter was to pay a rental of twenty pence. Evidently the abbot gave Walter the land Elias had once held, plus three more acres, and raised the rent.

The appearance of Abbot Richard's name does not date these donations precisely, because there were two abbots named Richard who ruled Grimsby Abbey in succession from perhaps as early as 1173 to about 1226. However, the names of some of the witnesses, especially Odo Galle, William of Karleton, William of Manby, and Alan of Grimoldy, appeared frequently in other late-twelfth-century documents.[3] The probability that Gilbert Fitz Roger owned the post-mill first suggests an even earlier date.

By the early thirteenth century, the Knights Templar were operating a windmill in the town of Grimsby and were trying to make all the local people use it.[4] The Cistercians of Meaux Abbey, across the Humber estuary in Holderness, also had a post-mill at Aylesby, just outside Grimsby, which they gave to the Wellow canons about 1300.[5]

1. For the original charter, see P.R.O., Ancient Charter D.L. 36/2, charter 113; briefly calendared in unpublished P.R.O. catalog, *Duchy of Lancaster, Cartae Miscellanae,*

2:54, charter 113. It was also transcribed in the beautiful cartulary, Duchy of Lancaster Coucher Book, MS. D.L. 42.2, fol. 390.

2. For the original charter, see P.R.O., Ancient Charter D.L. 36/2, charter 22; calendared in *Duchy of Lancaster, Cartae Miscellanae*, 2:41, charter 22; transcribed in MS. D.L. 42.2, fol. 405. This charter, the one cited in footnote 1, and the writs pertaining to mills at Silverley, Little Foxmold, and Evershaw are the earliest original writs to include mention of a post-mill.

3. On the Grimsby abbots, see Knowles, Brooke, and London, *Heads of Religious Houses*, p. 189. On the witnesses, see Frank M. Stenton, *History of the Danelaw*, esp. charters 539–43, 545–51, 554; and his *Free Peasantry*, charters 95, 103, 115, 122, 123, 130, 144, 153.

4. Edward Gillett, *A History of Grimsby*, p. 14. The site of the mill seems to have been a gift from William Fitz Ralph before 1185. This important sheriff may have been the donor of an early post-mill to Spalding Priory, also in Lincolnshire. The grant was first mentioned in 1263, in a general confirmation by Henry III; B.L. Additional MS. 35296 (Spalding Priory Cartulary), fol. 10.

5. B.L. Lansdowne MS. 424 (Meaux Abbey Cartulary), fol. 59; this was a grant from Abbot Roger of Meaux.

## 19. LINCOLNSHIRE: NORTH ELKINGTON   −1189 × 1194

Gilbert of Benniworth, who gave the nuns of North Ormesby a windmill at Friskney (see Appendix entry no. 16), also gave them twenty acres and a post-mill (*molendinum ad ventum*) in North Elkington. Both the original deed and its cartulary copy have been lost. The antiquarian Roger Dodsworth copied an extract from the charter from a since vanished register.[1] Dodsworth claimed that the charter pertaining to Gilbert's gift was dictated by William the son of Philip II of Kyme. This could be correct, but more likely the author was William the son of Philip I of Kyme (d. 1189 × 1194), Gilbert's lord and a man with whom he frequently attested other deeds. Many of Gilbert's other charters, some with personal seals, still exist.[2] The village of North Elkington might be of interest for archaeological study. It is now much smaller than it was in medieval times.

1. Oxford University, Bodleian Library Dodsworth MS. 135, fol. 147. Dodsworth copied this from fol. 90 of a lost Ormesby register.

2. Walter de Gray Birch, *Catalog of Seals in the Department of Manuscripts of the British Museum*, 2:248, items nos. 5685–94; B.L. Harley Charters 45 I 52, 45 I 53, 45 I 56, 45 I 77, 46 A 7. Some of these are printed in Frank M. Stenton, *History of the Danelaw*, charters 7, 32, 33, 34, 50, 51 (all for Bullington Priory) and 217 (for Newhouse Priory).

## 20. LINCOLNSHIRE: SWINESHEAD   −1168

Sometime before his death in 1168, William of Amundeville, steward of the bishop of Lincoln, gave to the Cistercian monks of Swineshead Abbey an unidentified plot, a rural tenement, and five strips of arable land where a windmill was to be raised (*ubi molendinum ad ventum situm fuerit*).

Although William's detailed charter has disappeared, the general terms of his grant are included in a long confirmation of the abbey's possessions issued by King Henry II between 1166 and 1179, most likely in April 1170. The king's charter has also vanished, but in 1316 it was quoted by King Edward II in a lengthy further confirmation of Swineshead privileges. Thus the evidence is a few hands removed from the actual event.[1]

The fact that the windmill site was small—only five strips—suggests that William of Amundeville had prior experience with post-mill planning. He may have let the abbey bear the construction costs, in which case he proved himself to be even more shrewd, if less generous. Several windmill donors made similar ambiguous arrangements. In addition, they usually restricted the acreage because a mill needed only a small space and because arable land was valuable. Locating the machines in the open fields eliminated some transportation costs, allowed mill builders to take advantage of high ground, and ensured the availability of unobstructed wind. Another popular decision was to locate a windmill beside an existing water mill.

The matter-of-fact tone of the charter recording William's donation indicates that by 1168 donors and scribes had accepted the construction of wind-powered machines as a common occurrence. The abbot of Swineshead at that time, Gilbert of Hoyland (d. 1172), was a learned preacher and self-conscious stylist who liked to use unusual graphic analogies. In his many sermons he often referred to the dread cold wind blowing in from the sea, but none of them included a windmill image.[2]

The monks of Swineshead were a rather independent, secularly minded group. The abbey did not have the best of reputations. Pope Alexander III (1159–1181), a stern reformer, chastised the monks for their ownership of serfs and other materialistic practices. Rumors would later circulate that King John was poisoned at Swineshead. Abbot Gilbert, another austere individual, oddly criticized the brethren who spent too much time reading.[3] It would be interesting to know whether William's engine was judged a concession to laxity or a distraction from things of the spirit.

William's career and avocational interests paralleled those of other windmill enthusiasts. For example, he often associated with Master Ralph de Beaumont, a famous physician who served the royal family and was a fellow member of the Lincoln bishop's entourage.[4] A knight and the second of five sons, William became steward of the bishop of Lincoln late in life, after his elder brother, Walter (a former steward, as was their father), had retired to Elsham Hospital. Their mother, Beatrice, a formidable woman, had established the hospital after her husband's death; she encouraged all her children to make it a family charity. William and his brothers regularly contributed to its upkeep. In addition to Swineshead, they helped support priories at Rufford in Nottingham and Swine in Yorkshire; William made gifts to the cathedrals at Durham and Lincoln.

Many of the family's gifts to churches were mills (generally water mills) or tithes on their profits.[5] On one occasion Walter gave Elsham Hospital three mills on the Trent River and a fourth away from the river in the nearby village of Winthorpe, Nottingham. It is tempting to assume that the latter was a windmill. When this grant was later confirmed by the bishop of Lincoln and attested by Ralph the physician, Thurstan the carpenter was leasing all four mills for an annual rent of twelve shillings.[6]

1. *Calendar of the Charter Rolls*, 3:318–19. King Edward's confirmation is printed in Dugdale, *Monasticon Anglicanum*, 1:173; that version refers to *ubi molendinum ad ventum situm fuerat* (rather than *fuerit*). Lynn White, Jr., thinks that windmills were not known before 1185 and that the document's language is unusual and is probably the result of a later marginal gloss; *Medieval Technology and Social Change*, p. 162. In light of the discovery of so many windmills, some of which were described as being similar to William's, there is no longer any reason to question the validity of this text or the early date for the existence of the Swineshead post-mill. Maurice Beresford and John K. St. Joseph (*Medieval England: An Aerial Survey*, p. 64) suggest a date between 1163 and 1181 for King Henry's confirmation, but it was certainly issued before 1179, when Richard de Lucy, one of the witnesses, died. R. W. Eyton (*Court, Household, and Itinerary of King Henry II*, p. 136), suggests April 1170.

2. For a discussion of his works, see *Gilbert of Hoyland*.

3. David Knowles, *The Monastic Order in England*, p. 351. See also Christopher R. Cheney, "English Cistercian Libraries: The First Century," in Cheney, *Medieval Texts and Studies*, p. 334; Dugdale, *Monasticon Anglicanum*, 5:337.

4. On Ralph's career, see Edward J. Kealey, *Medieval Medicus: A Social History of Anglo-Norman Medicine*, esp. pp. 143–46.

5. See Charles T. Clay, "The Family of Amundeville."

6. D. M. Smith, ed., *Episcopal Acta*, 1:73, document no. 112.

## 21. MIDDLESEX: CLERKENWELL   c. 1144

There are good reasons for assuming that the mill at Clerkenwell was indeed a windmill. About 1144 Jordan Bricett established a priory for Augustinian canonesses at Clerkenwell, just outside London, and granted to the nuns and to Robert their chaplain fourteen acres and a "place and site for making a mill" (*locum et sedem ad molendinum faciendum*).[1] His procedure was not unlike William of Amundeville's when he made his donation to Swineshead Abbey (Appendix entry no. 20).

Neither the extant cartulary copy of the charter recording Jordan's grant nor the copies of its confirmations refer specifically to a windmill, but as mentioned, the word *molendinum* was sometimes used interchangeably to mean windmill and water mill. The sixteenth-century antiquarian John Stow stated explicitly that in 1100 Jordan Bricett gave the canonesses a site upon which to erect a windmill.[2] This Elizabethan was wrong about the foundation date of the priory, but he could easily have had access to documents specifying that Jordan's plan was to build a post-mill. A windmill definitely stood in the nuns' field sometime before 1430.[3]

1. *Clerkenwell Cartulary*, pp. 30–31, charter 40. For discussion of this mill see Chapter 4.

2. John Stow, *A Survey of the Cities of London and Westminster and the Borough of Southwark*, 2:85.

3. Bennett and Elton, *History of Cornmilling*, 2:253.

## 22. NORFOLK: ATTELBOROUGH   –1198

In 1198 Ralph de Burg owned a windmill (*molendinum de vento*) at Wrogland, in Attelborough. There does not seem to be any way to estimate how old this mill was in 1198. It was cited as a landmark in a property dispute between William de Fossato and William de Arden and his wife, Isolda.[1]

The earls of Sussex, who were major patrons of Wymondham Priory, also held land in Wrogland.[2] The priory, five miles northeast of Attelborough, was the owner of several windmills.

1. *Feet of Fines, 10 Richard I, 1198–1199*, pp. 88–89, item no. 203; also *Feet of Fines, Norfolk*, ed. Rye, p. 23.

2. For information on the earls' holdings, see Dugdale, *Monasticon Anglicanum*, 6:467.

## 23. NORFOLK: BETWICK   c. 1200

Hugh de Albini was one of the many benefactors of Wymondham Priory. He was the younger son of Queen Adeliza and William de Albini, the first earl of Arundel (d. 1176). Hugh's brother, William, was also an earl (1176–1193), as was his nephew, William (1193–1221). Hugh gave Wymondham his windmill (*molendinum meum ad ventum*) at *Betewit*, as well as its site and adjoining two acres.[1] His brother gave the monks a post-mill (or, more probably, two of them) in Wymondham.[2] The date of Hugh's death is unclear; his gift could have been made as late as 1221 because the charter refers to an earl William, but this seems extremely unlikely.[3]

1. B.L. Cotton MS. Titus C VIII (Wymondham Priory Cartulary), fol. 64. The cartulary scribal heading refers to a grant of Earl Hugh de Albini, who was an earl from 1224 to 1243, but the text makes clear that this is an erroneous identification. The scribe made several other mistakes in this cartulary; one was repeated by Dugdale in *Monasticon Anglicanum*, 3:324. He also called the place both Barwick and Berwick, neither of which seems an equivalent of *Betewit*.

2. See Appendix entries nos. 34, 35.

3. Members of the de Albini family had some involvement with an unspecified mill called *Bicewille* before 1176, but this structure was probably in Sussex; *Boxgrove Cartulary*, p. 46, charter 56.

## 24. NORFOLK: BURLINGHAM   –1200

Jocelin of Burlingham gave Hickling Priory a post-mill and the half-acre on which it stood. His donation charter has disappeared, but Roger of Suding confirmed his late uncle's grant of the windmill (*molendinum venticum*) and the half-acre of arable land on which it once stood.[1] Evidently the canons had moved the windmill to a better location and returned the ground to cultivation. There are three Burlinghams in Norfolk (North Burlingham, Burlingham St. Peter, and South Burlingham); all are a few miles east of Norwich and south of Hickling. It is not clear which was the location of Jocelin's mill.

1. Oxford University, Bodleian Library Tanner MS. 425 (Hickling Priory Cartulary), fols. 31–31v. For a discussion of the Hickling post-mills, see Chapter 8.

### 25. NORFOLK: FINCHAM   −1200

Fincham is a small Norfolk village a short distance inland from the port of King's Lynn. It was one of the many places that were under the jurisdiction of several lords. This fragmentation weakened manorial authority and encouraged a relatively independent peasantry. Castle Acre Priory in Norfolk was the recipient of numerous small gifts. Philip of Burnham gave the monks a windmill in Fincham (*unam molendinum venti in Fincham*) with its attachments, as well as the whole strip of land on which it was located. He made this grant with the consent of his wife, Emma, and of his son, William. In the charter, Philip indicated that he had obtained the site of that mill (he used only the term *molendinum*) from William Fitz Osbert in exchange for five acres of land (quite a sizable area) in the same village. It would seem that Philip was eager to build a post-mill and wanted to find suitable land for it. His gift was witnessed by ten men from the area. Emma issued an almost identical charter, apparently at the same time, in which she stated that the exchange with William Fitz Osbert preceded Philip's grant.[1] Evidently, Philip and Emma were equal partners in this transaction.

It is difficult to determine when the windmill was constructed. The Castle Acre Priory Cartulary was compiled about 1260, and although the names of witnesses are usually specified, its charters are notoriously hard to date, partly because the monastery received endowments from many obscure individuals. Nevertheless, the cartulary yields a few facts about the Fincham patrons.

Emma was descended from a lesser branch of the L'Estrange family. Her first husband, William of Felsted, was the son of a rather unprincipled manager of monastic estates in Essex and Gloucestershire. After his death between 1192 and 1198, Emma married Philip of Burnham, who died about 1215. William of Burnham may have been Philip's son by another wife.[2] The names of Philip and his brother, Frederick of Hackford, who were both retainers of the Warenne family, frequently appear with that of Earl Hamelin (1168–1202).[3] In 1201 Philip courageously stood up in court and defended his rights to some wardships that Archbishop Hubert Walter coveted.[4] In 1214, Philip served as a guardian of Bury Saint Edmunds during a short abbatial vacancy.[5] He seems to have been ambitious and enterprising like so many of the windmill enthusiasts encountered in this study.

The Fincham post-mill probably began functioning long before 1215, because one of the signatories to the charter recording Philip's grant attested other documents between 1136 and 1153. Moreover, a considerable interval elapsed between Philip's donation and John Talbot's rental of the windmill (*molendinum suum ad ventum*) and its attachments from Prior Adam (1239–1260) for the small sum of four pennies a year.[6] Talbot's statement refers to William of Burnham, Philip's son, in the past tense. It thus seems reasonable to conclude that the Fincham installation was raised before 1200.

In 1222 a William of Burnham farmed the manor of Beauchamp, in Essex, for the canons of Saint Paul's Cathedral. That estate also boasted a post-mill that was said to be capable of returning a profit of one mark a year.[7] Burnham was a fairly common name, however, and it is not certain that this William was Philip's son.

1. B.L. Harley MS. 2110 (Castle Acre Priory Cartulary), fol. 85.

2. Marjorie Chibnall, *Charters and Custumals of the Abbey of Holy Trinity, Caen*, pp. xlii, 4.

3. On Hamelin, Philip, and Frederick, see B.L. Harley MS. 2110, fols. 6–6v. On Philip and a mill at Witewell, and on Frederick and a water mill (*molendinum ad aquam*) at Hackford, see *Feet of Fines, Norfolk*, Pipe Roll Society Publications 70:123, item no. 252 (for the year 1210).

4. Young, *Hubert Walter*, p. 156.

5. R. M. Thomson, ed., *The Chronicle and Election of Hugh, Abbot of Bury St. Edmunds and Later Bishop of Ely*, pp. 80–81.

6. B.L. Harley MS. 2110, fol. 85.

7. *Domesday of St. Paul's*, p. 28.

## 26. NORFOLK: FLOCKTHORPE   −1198

In the late twelfth century, Stephen of Camais gave the monks of Wymondham Priory a windmill in Flockthorpe (*molendinum meum ad ventum de Flokestorp*) with attachments, as well as its site. He reserved milling rights for his family; the tithe of the mill went to the Church of Saint George in Hardingham.[1]

Stephen was the son of Robert, the son of Humphrey, who was a constable of the earls of Hertford and a benefactor of the family's favored religious house, Stoke-by-Clare Priory in Suffolk.[2] Walter Fitz Robert also endowed the priory, as did Stephen.³ Stephen went on the Third Crusade, returned to England, and died before 1198, when his son, Ralph, was still a minor.⁴ Stephen's cousins controlled the village of Steeple Bumpstead in Essex, which also had a windmill before 1199–1226.⁵

1. B.L. Cotton MS. Titus C VIII (Wymondham Priory Cartulary), fol. 66.

2. *Stoke-by-Clare Cartulary*, charters 328, 330. For appearances of Robert Fitz Humphrey's name see charters 23, 28, 30, 33, 34, 36, 38, 43, 46, 49, 315, 316, 365, 366. He seems to have died in the 1180s, certainly after 1173.

3. On Walter Fitz Robert's gift, see ibid., charters 183, 269. For information regarding Stephen's gifts, see charters 99, 326, 329, 622.

4. *Pipe Rolls, 1 Richard I*, p. 46; *3 Richard I*, p. 44; *10 Richard I*, p. 107. Ralph confirmed some of Stephen's gifts; he died in 1259; *Stoke-by-Clare Cartulary*, charter 327.

5. The post-mill was granted by John Fitz Lambert (known to have lived 1198–1243) to Hugh the Chamberlain (known in 1221), who later gave it to Stoke-by-Clare Priory. John's grant was attested by Walter Fitz Humphrey, Stephen's second cousin (whose name occurs between 1199 and 1226), *Stoke-by-Clare Cartulary*, charters 323, 324.

## 27. NORFOLK: HEMPNALL    1136–1198

Before his death in 1198, Walter Fitz Robert, steward of the king, gave the Abbey of Bury Saint Edmunds half an acre on his manor of Hempnall, at a place called Longbridge, on which to erect a windmill for their use (*ad faciendum ibi molendinum ad ventum ad opum suum*).¹ The gift was probably made between 1170 and 1198, although it could have been made at any time during the baron's long career. The monks built the mill shortly after they accepted the land. Walter is always credited with giving them both the post-mill and its site.

1. C.U.L. MS. Mm. 4.19 (Black Book of Bury), fol. 164. For a detailed discussion, see Chapter 5.

## 28. NORFOLK: HERRINGBY    c. 1200

The Herringby post-mill was one of the windmills belonging to Hickling Priory. It came to the monastery in two donations. Walter of Herringby gave his windmill (*molendinum meum venticum*) in that village and all its appurtenances to the convent.¹ However, at that time, he could give only half the land on which it stood. John Fitz Hodbert was then renting the other half but eventually gave it to the priory also.²

1. Oxford University, Bodleian Library Tanner MS. 425 (Hickling Priory Cartulary), fol. 36.

2. Ibid., fol. 23. Walter of Herringby was involved in litigation in 1208 and 1209; Doris M. Stenton, *Pleas before the King*, 4:105, item no. 3454; 4:206, item no. 4177.

## 29. NORFOLK: HICKLING   c. 1185

The Norfolk Broads was an unlikely area for a priory, but in 1185 Theobald of Valeins II selected twelve acres in Hickling and invited four Augustinian canons from Saint Osyth's Priory in Essex to establish a monastery. Thereafter, he made additional small grants of land, rents from his properties, and churches. Theobald's charters were usually quite specific. For example, when he gave the canons the site of their windmill (*sedem cuisdam molendini ad ventum*) in Hickling, he indicated that the plot, in his field called Eastwihig (?), was 100 feet in length by 100 feet in width. He also recorded that the distance from the mill to the road, called Bladegate,[1] was 12 feet. Evidently, he had given them the mill before he donated the land beneath it.

There were several windmills in Hickling, but the others cannot be described exactly because much of the mid-thirteenth-century Hickling Priory Cartulary is difficult to read. Both the text and the microfilm copy of the first page are completely illegible. The parchment is uncommonly dark and something that was spilled on the initial folio obliterated many crucial words, including the name of the person who issued this long confirmation. In at least one instance, however, this grant refers to more than one windmill. Several relevant words are unclear, including the donor's name and the mills' location.[2] They were probably constructed at Hickling, but they may have been at Palling, the last place name in the text that is legible.

There were certainly at least two post-mills at Hickling by the mid-thirteenth century. During the tenure of Prior Nicholas (1232–1248), the convent was forced to seek payment of the tithes on the profits of one windmill (*molendinum ad ventum*) from Sir Brian of Hickling and his wife, Julia.[3] This must have been a different windmill from the one Theobald gave the canons, for theirs need not have been tithed. Only one of the Hickling post-mills can be dated before 1200. Brian's mill might be much older, however, since his name appears elsewhere before 1200.[4]

1. Oxford University, Bodleian Library, Tanner MS. 425 (Hickling Priory Cartu-lary), fol. 42v. For more on Theobald, see Richard Mortimer, "The Family of Rannulf de Glanville," pp. 7–12; *Leiston Cartulary*, p. 66, charter 15.

2. Bodleian Library Tanner MS. 425, fol. 9.

3. Ibid., fol. 1.

4. *Leiston Cartulary*, p. 68, charter 19; C.U.L. MS. Mm. 2.20 (Bromholm Priory Cartulary), fol. 49v. Brian was the son of Geoffrey of Hickling; *Leiston Cartulary*, p. 151, charter 146. He had a brother named Robert; C.U.L. MS. Mm. 2.20, fol. 3. This family seems to have been connected with the Glanvilles. A Geoffrey owned a mill (post-mill?) in Hickling during Theobald's tenure; Bodleian Library Tanner MS. 425, fol. 43v. Robert Fitz John, who may have known this Hickling family, gave a windmill at Brotford in Norfolk to Bromholm Priory before 1229; C.U.L. MS. Mm. 2.20, fol. 3. Geoffrey of Hickling, a knight, was involved in litigation in 1209; Doris M. Stenton, *Pleas before the King*, 4:213, item no. 4200.

## 30. NORFOLK: PALLING   −1200

The Palling windmill is one of those structures which can easily be overlooked because it was not consistently identified as a wind-powered machine. Such indefinite terminology suggests that it was an old installation. Bartholomew of Beighton (*Beilham*) was one of the principal benefactors of Hickling Priory. He made a series of modest, but cumulatively impressive, donations. One gift was the site of a mill (*sedem molendini*) at Palling and the franchise over his men there and in Waxham.[1]

Later the land passed to Simon Bigot, who returned it, except for a small piece called Millwog (*milne Wog*),[2] to Hickling Priory. He referred to it as the windmill site (*in sede molendini ventici*) the canons had been given by Bartholomew of Beighton. Thus Bartholomew's gift was a windmill site.

The date of these gifts can only be estimated. Bartholomew was a nephew of Thomas of Beighton, whose name appeared between about 1170 and 1187.[3] Simon Bigot appeared in Norfolk documents between 1130 and 1140.[4] The spelling of the word "windmill" depended on the cartulary scribe's unusual preference. The form used for Waxham (*Wax-thuesham*) was antiquated. Thus the mill probably existed long before 1200.

1. Oxford University, Bodleian Library Tanner MS. 425 (Hickling Priory Cartu-lary), fol. 24. This charter was copied a second time in the Hickling Priory Cartulary, fol. 29.

2. Bodleian Library Tanner MS. 425, fol. 33. Simon also issued a general confirmation of Bartholomew's gifts (fols. 24–24v).

3. Oxford University, Bodleian Library, Suffolk Charter 239; H. O. Coxe and W. H. Turner, eds., *Calendar of Charters and Rolls Preserved in the Bodleian Library*, p. 536; R. M. Thomson, "Twelfth-Century Documents from Bury St. Edmunds," p. 818, charter 29.

4. Lewis C. Loyd and Doris M. Stenton, eds., *Sir Christopher Hatton's Book of Seals*, p. 196, charter 284.

## 31. NORFOLK: ROLLESBY  c. 1200

Women were important windmill donors. With the consent of her lord (and husband?), Walter of Somerton, Beatrice of Somerton gave Hickling Priory her windmill with its site and appurtenances (*molendinum meum venticum cum sede sua et pertinentibus*) in the village of Rollesby.[1] Her charter located the mill quite precisely, "standing at Tokesgap near the road which extends from the bridge of Bastwick up to the bridge of Burgh (St. Margaret)." Her gift was made for the salvation of her soul and the souls of her relatives. She stipulated that one penny from the mill's profits was to be rendered to her lord each year at the Feast of Saint Edmund. The charter cannot be dated with certainty before 1250, the date of the cartulary, but its antiquated style and the generosity of its terms suggest that it was probably issued before 1200.

1. Oxford University, Bodleian Library Tanner MS. 425 (Hickling Priory Cartulary), fol. 33v.

## 32. NORFOLK: SNETTISHAM  −1200

Roger of Rusteni was a vassal of the earls of Arundel, the patrons of Wymondham Priory. He served the de Albini (sometimes written d'Aubigny) family for more than seventy years. Roger was also a generous benefactor of the Wymondham monks. Before 1182 he gave the priory a marsh in Snettisham, near The Wash.[1] After 1193 he donated forty acres at Sharnbourn.[2]

Roger was an enthusiastic windmill advocate. Between 1193 and 1199 he attested Earl William III's long confirmation of priory grants, including its post-mills.[3] He gave the monks his windmill (*molendinum meum ad ventum*) at Snettisham, as well as the three small parcels of land on which it was located.[4] Roger stated in the charter that this gift was to commemorate his death, that of his wife, and that of the earl. He further directed that each year, thirteen poor people were to be fed on the anniversary of his death. Roger's donation was thus another link between social welfare concern and technological innovation. His grant cannot be dated precisely, but it was probably made long before 1200, both because of his longevity and because his son, Nigel, was making grants of his own as early as 1176 × 1193.[5]

1. B.L. Cotton MS. Titus C VIII (Wymondham Priory Cartulary), fol. 49. Roger's charter (1193–1221) seems to be a reconfirmation, for in 1182 Bishop John of Norwich mentioned that the monks had already received the marsh (fol. 10). Earl William III (1193–1221) also confirmed Roger's grant (fol. 20).

2. Ibid., fol. 48v; Dugdale (*Monasticon Anglicanum*, 3:324) erroneously claimed that this was a gift of Robert of Rusteni.

3. B.L. Cotton MS. Titus C VIII, fols. 19v–20.

4. Ibid., fol. 48. There were other mills at Snettisham, but it is not known whether they were windmills; see fol. 46v. See also *Feet of Fines, Norfolk*, Pipe Roll Society Publications 70:33, item no. 72 (for the year 1205).

5. B.L. Cotton MS. Titus C VIII, fol. 18.

## 33. NORFOLK: WAXHAM   c. 1190

Thomas of Thurne (*Tyrne*) was one of the earliest and most generous benefactors of Hickling Priory. His wife also made several grants. One of his gifts to the Augustinian canons included half of the mill in Waxham (near the North Sea), the mill site, the strips of land pertaining to the mill, and one acre of arable land next to it.[1] The other half of the mill belonged to Warin of Rollesby, who donated it separately. That grant was not recorded in the priory's cartulary, but a subsequent agreement between Thomas and the priory about those lands was preserved.[2] It is interesting that Thomas and Warin cooperatively managed one mill.

Not until later did the records indicate that the facility donated to Hickling Priory was a post-mill. Thomas of Belham (presumably an alternate spelling of Beighton), who was the son of Walter and the nephew of Thomas of Thurne, confirmed all his uncle's grants, including the canons' windmill (*molendinum venticum*) in Waxham, with the mill site and the land adjoining it.[3]

The names of Warin of Rollesby and his wife, Agnes, appeared in a charter of Bartholomew of Martham written about 1190.[4] The name of Thomas of Beighton appeared between about 1170 and 1187.[5] The situation in Waxham was somewhat like that in Palling. A number of the individuals in Waxham with whom Thomas of Thurne associated had Scandinavian names, reminiscent of Danelaw times. Such a person was Gunnar the carpenter, who lived next to Thomas.[6] He might well have been a windmill builder.

1. Oxford University, Bodleian Library Tanner MS. 425 (Hickling Priory Cartulary), fol. 24v. Thomas is still mentioned in 1209; *Feet of Fines, Norfolk*, Pipe Roll Society Publications 70:95–96, item no. 200.
2. Bodleian Library Tanner MS. 425, fol. 2v.
3. Ibid., fol. 1v. Thomas of Belham was known in 1223–1224.
4. Oxford University, Bodleian Library, Norfolk Charter 424; Coxe and Turner, *Calendar of Charters*, p. 211. Agnes also endowed Bromholm Priory; C.U.L. MS. Mm. 2.20, fol. 3.
5. See Appendix entry no. 30.
6. Bodleian Library Tanner MS. 425, fol. 6.

## 34–35. NORFOLK: WYMONDHAM   −1193

Wymondham Priory was established in 1107 by William de Albini, the cupbearer of King Henry I. William's son married the king's widow and eventually became an earl. The de Albinis are a confusing family because five consecutive generations of males between 1100 and 1224 called themselves William de Albini. Several of them died relatively young. Thus there was a rapid succession of earls who variously entitled themselves earls of Arundel, earls of Chichester, or earls of Sussex. The family's base was in the South at Arundel Castle, but they had extensive holdings in Norfolk. Like several other magnates, these earls believed that mills of all sorts made appropriate gifts for monastic communities, and they became enthusiastic post-mill supporters.

The first William gave the Benedictine monks at Wymondham at least four mills, two of which were clearly water mills. William, or his son, subsequently gave two more.[1] As frequently happened, once the initial enthusiasm had subsided, quarrels arose concerning the tithes of some of these mills. One such controversy was evidently not settled until 1230.[2]

The monks sometimes had difficulty gaining access to their mills, since they were on manorial property. King Henry II once sent the first earl, William (1139/41–1176), a stern note ordering him to grant the religious free entrance and exit. Henry declared, as he and his predecessors had in other writs, that he did not want to hear any more about this matter. He threatened to send the sheriff if the earl did not solve the problem.[3] To prevent such difficulties, many post-mill owners located their machines near roads, or specifically guaranteed free access to all customers. This ability to choose their site was an advantage not enjoyed by water mill operators.

The third William, who was the second earl from 1176, or 1186, to 1193, continued the family tradition of donating mills. Sometime before 1190 (that is, while he was still styled the earl of Sussex, rather than the earl of Arundel, the title he later used), this William gave the Wymondham monks at least two windmills in Wymondham. One charter had a proprietorial tone; it referred to a grant to "my monks" of the church of Wymondham. The text specified the mill location as Wymondham, but the scribal heading specified Eastfield. (Many of these early post-mills were raised on a lord's easternmost holding, perhaps to catch the stronger coastal breezes.)

William declared that he gave the windmill (*molendinum meum ad ventum*) in free alms in order that the anniversary of the death of his father and of his own death would be honorably commemorated each year. He also granted the mill site and franchise, except for those rights pertaining to his water mill (*molendini mei ad aquam*) in the same village when it was able to operate. He also quitclaimed to the monks their accustomed fishing rights in the millpond.[4]

Between 1193 and 1199, the third earl confirmed his ancestors' many gifts to Wymondham Priory. He seemed unsure which had been donated by his father and which by his grandfather. At the end of the document, almost as an afterthought, he mentioned the windmills (*molendina sua ad ventum*) in Wymondham. This indicates that at least two windmills had been granted before 1193. The lengthy confirmation was attested by Roger of Rusteni, who gave the monks a post-mill at Snettisham.[5]

The two windmills could have been donated by the first earl (that is, between 1139/41 and 1176), or they could have been donated early in the second earl's tenure. One mill could have been given by each donor. Alternatively, Earl William II (1176/86–1193) could have given one, as mentioned in the charter, and thereafter he may have given another. Or one mill could have been donated by Earl William II and one by Earl William III (1193–1221). In any case, it seems obvious they were twelfth-century post-mills and probably old ones at that. The village of Wymondham therefore had at least five mills: two water mills that William I gave to the priory, a third water mill that the earls used for their own grinding, and two post-mills. One monastic post-mill was to be used as auxiliary power, perhaps in winter, for the earl specifically ordered that the rights of his water mill were to be safeguarded.

In view of the Wymondham monks' interest in mills and in new types of mills, it is perhaps not surprising that the priory's cartulary began with copies of a few papal letters, including one to the Abbey of Saint Albans, Wymondham's motherhouse, from Pope Celestine III, who had decreed that windmill owners must pay tithes. This letter also concerned tithes, but this time Celestine referred to new mills (*de novis molendinis*).[6]

Besides the mills at Wymondham and the mill at Snettisham, the priory owned post-mills at Betwick (a gift from the second earl's brother, Hugh), and at Flockthorpe. There was also a post-mill at Forncett, nearby. It was not mentioned until 1279, when it was described as debilitated, probably indicating that it was quite old.[7]

1. B.L. Cotton MS. Titus C VIII (Wymondham Priory Cartulary), fols. 15, 18, 22, 36.

2. Ibid., fols. 110–12.

3. Ibid., fol. 36.

4. Ibid., fol. 21.

5. Ibid., fols. 19v–20. This grant was made during King Richard's reign (1189–1199) by the son of the earl of Sussex and the grandson of the earl of Arundel, that is, Earl William II (1193–1221)—hence the dates 1193 and 1199. The windmills were not mentioned in the different royal confirmations of Wymondham's possessions, but these charters were not always detailed.

6. Ibid., fol. 14.

7. Frances G. Davenport, *The Economic Development of a Norfolk Manor, 1086–1565*, p. 36. On the post-mill at Betwick, see Appendix entry no. 23; on the Flockthorpe mill, see entry no. 26.

## 36. NORTHAMPTONSHIRE: BRAMPTON ASH   −1190

Robert Fitz Adam of Brampton gave the monks of Saint Mary of Pipewell, a Cistercian abbey founded in 1143, the site of a windmill (*sedem molendini ad ventum*) next to the vill of Brampton Ash (*Bramtone*) and the ditch that surrounds that site. The land was said to abut the road from Brampton Ash to Dingley. Robert gave it to the monks in perpetual free alms, with license to construct a mill (unspecified) there.[1] This probably means that a post-mill had not yet been erected, but

Robert could also be giving the monks permission to build a second mill.

This charter, which contains the earliest mention of a ditch surrounding a post-mill, was recorded in the Pipewell Abbey Cartulary along with numerous mid-twelfth-century documents, including one from Richard Turc, master of the Hospitallers (1165–1170), and several from Bishop Robert of Lincoln (1148–1166). The likelihood is therefore strong that Robert's deed is also a midcentury document. A subsequent notation in the margin indicated that this windmill yielded twelve pence a year. Brampton Ash is a village not far from Wigston Magna, the location of an early post-mill.

1. B.L. Additional MS. 37022 (Pipewell Abbey Cartulary), fol. 10v. No witnesses were recorded.

## 37. NORTHAMPTONSHIRE: DAVENTRY   –1158

Matilda of Senliz's gifts to Daventry Priory before 1158 included three unspecified mills. Her descendants regularly confirmed these gifts. Almost a century later, in 1244, Roger the miller negotiated an agreement about three Daventry Priory mills, which were then identified as two water mills and a windmill.[1] They must be the same three installations mentioned decades earlier, for only three mills had been referred to consistently during the intervening eighty-six years.

1. B.L. Cotton MS. Claudius D XII (Daventry Priory Cartulary), fols. 5–6v; *Christ Church Cartulary*, pp. 54–55. For a detailed discussion of this post-mill, see Chapter 5.

## 38. NORTHAMPTONSHIRE: DRAYTON   –1164 × 1197

When Matilda of Senliz donated three mills and other gifts to Daventry Priory, she insisted that no other mills were to be built in the area unless they were to be used for the work of the monks. Walter Fitz Robert confirmed his mother's grant between 1164 and 1197. However, William of Drayton erected an unspecified mill in defiance of their proscription. Walter forced his knight to surrender the offending facility to him and subsequently granted it to the monks. In 1244, it was clearly identified as a post-mill. It was still working between 1289 and 1318 and was then called the Ryen Hill Mill.[1]

1. B.L. Cotton MS. Claudius D XII (Daventry Priory Cartulary), fols. 5–7; *Christ Church Cartulary*, pp. 53–56. For a detailed discussion of this windmill, see Chapter 5.

## 39–40. NORTHUMBERLAND: NEWCASTLE   –1196

Thomas Bigod and his wife, Agnes de Glanville, claimed shares in several properties in and around Newcastle. Included were two wind-mills (*de duobus [mo]lend . . . venticiis*) beyond the eastern part of the town that were held by Robert of Peasenhall, and the acre of land on which they were built. The whole property was a fief of William de Bikere. By the terms of a final concord reached at Westminster on May 20, 1196, Thomas and Agnes surrendered all their claims to Robert in return for a hawk valued at one mark.[1]

Robert of Peasenhall had other holdings in the area. From 1195 to 1198 he owed the Exchequer the scutage tax of one mark for a city tenement.[2] He was well known among windmill promoters. Between 1189 and 1192, for example, he appeared with Burchard (the treasurer of York Minster) and with a member of the large Amundeville family.[3] Robert's home base, Peasenhall in Suffolk, was not far from a number of post-mills. Newcastle was particularly receptive to mills; its town charter specifically provided that each burgess could have his own oven and a mill.[4] Robert may have exceeded the quota with regard to the latter.

1. *Feet of Fines, 7–8 Richard I, 1196–1197*, p. 1.
2. *Pipe Rolls, 7 Richard I*, p. 25; *8 Richard I*, p. 93; *9 Richard I*, p. 10; *10 Richard I*, p. 144.
3. William Farrar and Charles T. Clay, *Early Yorkshire Charters*, 9:143, charter 71 (1189–1192).
4. There are two slightly different versions of the town charter; one refers to mills in general, the other to hand mills. See Alphonsus Ballard, ed., *British Borough Charters, 1042–1216*, pp. 96–98.

## 41. OXFORDSHIRE: CLEYDON   –1196

Henry d'Oilli, constable of the king, allowed the Abbey of Saint Mary of Oseney free entrance to and exit across his land at Cleydon in order to use his windmill (*ad molendinum suum ad ventum*).[1] A wise pro-moter, he did not donate the mill but rather encouraged the canons to use it. The manor of Cleydon had been a gift from King Henry I when one of his mistresses, Edith Forne, the mother of Robert Fitz Roy, married into the d'Oilli family.

This post-mill and the charter recording its donation are difficult to date precisely. One reason is that there are three possible authors of the charter: Henry d'Oilli I (d. 1163), Henry d'Oilli II (d. 1196), and Henry d'Oilli III (d. 1232). Moreover, editorial estimates of the charter's date differ. The heading of the medieval copy in the Oseney Abbey Cartulary identifies the donation as that of Henry d'Oilli II. This seems reliable, since the next few charters in the collection date before 1189. The modern editor postulated a date of 1205 for the charter, but offered no reason for his assumption. Unfortunately, no witnesses' names were recorded.

There were several unspecified mills on the d'Oilli properties. Robert d'Oilli (d. 1142), the original patron of the abbey, gave the canons hunting and fishing rights on the land next to Oxford Castle where his mills were located and the tithes levied on his land and mills in Oxford. He also gave them the Church of Cleydon and the entire district of Eton, as well as its mill. These grants were confirmed by his son, Henry d'Oilli I.[2] There is no way to determine whether any of the relevant documents were referring specifically to windmills. However, it should be remembered that after the death of Henry d'Oilli I his widow, Matilda de Bohun, married Walter Fitz Robert, the post-mill enthusiast.

1. *Oseney Cartulary*, 5:209, charter 692.
2. Smith, *Episcopal Acta*, 1:34, document no. 53; 1:132, document no. 213.

## 42. SUFFOLK: DUNWICH  −1199

In 1185 the Knights Templar undertook a comprehensive survey of their English lands and possessions that explicitly identified one windmill at Weedley in Yorkshire. At the very end of the written tabulation a second twelfth-century scribe made an entry that recorded a gift made by an unnamed king to the Templars of a plot of land (*messuage*) and a windmill (*unum molendinum vento*, an unusual Latin form) in Dunwich. Both had been leased by a tenant, John of Cove, for a rent of one-half mark a year, a rather large sum.[1] (John of Cove, the son of Thorald, was still alive in the early thirteenth century.[2]) This is the only entry pertaining to East Anglia in the entire survey; it was clearly an afterthought.

Unfortunately, the royal donor was not named. He was probably Henry II or Richard, because by 1199 most of Dunwich had been confirmed to the Templars by King Richard and King John.

Dunwich was once a thriving port but it has been largely swallowed up by the sea. Nearby Woodbridge had an early tidal mill, which still functions, and other early post-mills were located in the vicinity of Dunwich. The Augustinian canons of Blythburgh had one at their priory and Roger de Chesney gave them another, in Darsham.[3] The Chesney family was active in Cove.[4] In addition, Walter Fitz Robert had retainers in Dunwich.[5]

1. Beatrice A. Lees, ed., *Records of the Templars in the Twelfth Century: The Inquest of 1185*, p. 135.

2. *Blythburgh Cartulary*, 2:190, charters 364–65.

3. Ibid., 1:38, charter 16 (–1250). On the post-mill at Darsham, see p. 127, charter 223, and p. 134, charter 237 (both charters probably dated 1218, or earlier).

4. Ibid., 1:104, charter 175 (–1199).

5. For example, Robert the son of Ulf of Dunwich. See ibid., 2:162–63, charters 300–302; Loyd and Stenton, *Hatton's Book of Seals*, p. 211, charter 305.

## 43. SUFFOLK: ELVEDEN   –1200

Although Samson, abbot of Bury Saint Edmunds (1186–1211), opposed Dean Herbert's effort to build a windmill at Haberdun (see the next entry), he was not opposed to the utilization of wind power. His abbey operated several post-mills.

Before 1200, with the consent of the abbey, Samson leased to Solomon of Whepstead the fully stocked manors of Ingham and Elveden. Their contract included an inventory of the animals and equipment on the manors and a list of the annual rents. Elveden was expected to yield the abbey sufficient food for two weeks each year. The equipment included two ratable water mills (*molendina ad aquam pacabilia*) and a good windmill with all its apparatus (*molendinum unum bonum ad ventum cum toto apparatu*).[1] The term "good" can be variously interpreted. It could refer to working order, implying that some local post-mills were inefficient. It could mean large, well constructed, or perhaps profitable. It might suggest legality (that is, the windmill was licensed).

Samson was in the process of eliminating abbey debts. Elveden manor had previously been controlled by the monastic cellarer; however, Solomon of Whepstead had been leased other abbey holdings and was evidently considered a good tenant. Lifetime leases such as Solomon's usually did not make allowance for the effects of inflation, and long-term tenants were difficult to control. However, his rent was to be paid in kind (produce), not money; moreover, Solomon may have been advanced in years at the time of the transaction. Abbot Samson generally shifted to limited-term contracts as he gained experience. Therefore, this contract with Solomon may have been executed early in the abbot's tenure. It was reinforced by an agreement reached in the king's court in Michaelmas 1202.[2]

1. Samson, *Kalendar*, p. 119, charter 77; copied from C.U.L. Additional MS. 4220 (Bury Cellarer's Cartulary), fol. 80v.

2. Samson, *Kalendar*, p. 119, charter 77; see also P.R.O. MS. D.L. 42.5 (Bury Cellarer's Cartulary), fol. 29v.

## 44. SUFFOLK: HABERDUN    c. 1182

About 1182 Dean Herbert erected a post-mill (*molendinum ad ventum*) at Haberdun, on the outskirts of Bury Saint Edmunds. Abbot Samson furiously demanded that it be torn down, but Herbert defended his right to utilize "the free benefit of the wind."[1]

1. Jocelin, *Chronicle*, pp. 59–60. For an extended analysis of this confrontation, see Chapter 6.

## 45. SUFFOLK: HIENHEL    −1198

An entry in the court records for October 18, 1198, states that Simon of Farnham proved that Ranulf Dun held half a windmill (*molendinum ad ventum*) from him in Hienhel, Suffolk, by hereditary right.[1] (I have not been able to identify Hienhel precisely; many villages in the shire have similar names.) Ranulf and his heirs were to rent the mill for six pence (a low rental), payable on Michaelmas, October 29. To obtain Ranulf's consent to the agreement, Simon gave Ranulf forty shillings.

Presumably Ranulf Dun had claimed that he owed no service to Simon. The dispute might have originated in an earlier generation. If Ranulf's service had been recent, local people would have remembered it. Thus, the mill probably existed long before 1198. One of the judges in the case, R(ichard Barre), archdeacon of Normandy, had acted as a deputy for Pope Celestine III (1191–1198) in a windmill case.[2]

1. *Feet of Fines, 10 Richard I, 1198–1199*, p. 22, item no. 32.
2. See Appendix entry no. 4.

## 46. SUFFOLK: RISBY  −1200

At least one early windmill (*molendinum ad ventum*) was located in Risby, which is four miles west of the Abbey of Bury Saint Edmunds. R——, the son of Ralf le Breton, granted it to the abbey sometime before 1200.[1] Although the charter is undated, it mentions Richard Fitz Maurice, who was active in 1166, as was Ralf le Breton.[2] The transaction was witnessed by Sir Richard of Ickworth, who attested a number of Abbot Samson's acts, including his establishment of Saint Saviour's Hospital in 1192.[3] Such windmill enthusiasts as Master Stephen, the son of Dean Herbert, and Solomon of Whepstead also attested the charter.

It should be noted that in 1203 Robert le Bretun and his son Roger reached an agreement about some of their lands in Wortham, Palgrave, and Burgate. Wortham, a village about twenty miles east of Risby, had a windmill (*molendinum venti*) at that time, but it was probably constructed much earlier.[4] Bretun (or Brito) was a common surname, so one cannot be positive that le Bretun and le Breton were the same family, but such a possibility seems likely. If they were, R(obert) le Bretun was probably a promoter who owned multiple post-mills—at least one in Risby and one in Wortham.

1. P.R.O. MS. D.L. 42.5 (Bury Cellarer's Cartulary), fol. 42v.
2. On Richard Fitz Maurice, see *Red Book of the Exchequer*, 2:395; on Ralph, see vol. 1, p. 66 (1186–1187) and vol. 2, pp. 358 (1166) and 503 (1210, perhaps an appearance by Ralph's son).
3. Jocelin, *Chronicle*, p. 121. For other appearances of Sir Richard's name, see Samson, *Kalendar*, pp. 76, 90, 102, 110; B.L. Additional MS. 14847 (White Book of Bury), fols. 69v–70. In 1200 Richard of Ickworth owed Bury Saint Edmunds the service of two knights; Jocelin, *Chronicle*, p. 121.
4. *Feet of Fines, Suffolk*, p. 200, item no. 419.

## 47. SUFFOLK: TIMWORTH  c. 1200

About 1200 Reginald of Groton and his wife, Amicia, gave the Abbey of Bury Saint Edmunds their windmill (*molendinum nostrum ad ventum*) in Timworth with its site and legal rights. Apparently they had considered themselves joint owners of the mill. Their grant was attested by R—— of Norton and Thomas le Jewich.[1]

Timworth, now called Timworth Green, is four miles north of Bury and was sometimes considered part of Ingham, one of the abbey's large manors. The scribal heading for the cartulary copy of Reginald and Amicia's charter was "Ingham." A post-mill in Tostock, the next village south of Norton, was owned by William, the son of Roger of Norton, perhaps the same R—— of Norton who witnessed the Timworth donation.[2]

1. P.R.O. MS. D.L. 42.5 (Bury Cellarer's Cartulary), fol. 44v.
2. See Appendix entry no. 48.

## 48. SUFFOLK: TOSTOCK    c. 1200

The post-mill at Tostock experienced rapid tenurial change—from lay sponsorship, to monastic control, and back to lay direction. The earliest known proprietor was an almost totally obscure individual, William the son of Roger of Norton. About 1200, or earlier, he granted his windmill (*molendinum meum ad ventum*) and its site in Tostock (*Thotestok*) to Sir Robert Danmartin, the marshal of Boulogne. Robert received the mill for his faithful service and for the large cash consideration of four and one-half silver marks. He could assign it to whomever he wished, but William was always to receive a penny a year as rent, a nominal sum yet enough to signify his ultimate ownership. Robert served William in some capacity, but this transaction was evidently still largely commercial in nature.[1]

In a contemporaneous agreement, William Fitz Roger also transferred to Sir Robert Danmartin a certain William, the son of Wilmer (*Wlmer*) of Tostock, and his tenement. This particular William seems to have been a serf, perhaps a miller. For the grant, Robert the Marshal gave William Fitz Roger of Norton five and one-half marks, an encouraging indication that men were valued more highly than machines. Furthermore, Robert was reminded that Wilmer's son was annually to render six half-pennies to William Fitz Roger.[2]

Robert Danmartin was one of the estate manager types who keep cropping up in this tale. By 1200 he was an attorney for the count of Boulogne and his wife. Most likely, he was also their steward and a relative of the count. Robert enlarged his holdings over the years and became a benefactor of Sibton Abbey in Suffolk. He did not enjoy uninterrupted bliss, however, for in 1213 he had to give the Exchequer one good horse in order to regain his wife, who had been captured by Eustace the monk. This ingenious master pirate, an ally of King John from 1205 and an admiral for the French after 1212, was the scourge of the Channel. A renegade in every sense, Eustace was even said to exercise demonic powers. The assault on Robert's wife probably reflects an old personal grudge, since Eustace had himself once been seneschal to the count of Boulogne. Ransom for the lady Danmartin was evidently first paid by the crown, which was still seeking reimbursement in 1230.[3]

About 1212, perhaps as supplication for his wife's deliverance, Robert the Marshal gave the Cistercian monks of Sibton, in free alms, his Tostock windmill (*molendinum meum ad ventum de Thotestok*), its site, and William the son of Wilmer. Danmartin declared that he held these from William Fitz Roger of Norton and that it was understood that the Sibton brothers would pay William nine pence a year.[4] The remark that the gift was granted in free alms was not a mere formality, for Cistercian regulations provided that those were the only forms of donation which the monks should accept.

William Fitz Roger confirmed the marshal's tripartite gift.[5] Sometime thereafter, he released the monks from the requirement that they pay him three pence each year for William the son of Wilmer and his tenement.[6] It is not clear whether this sum was part of the nine pence just mentioned.

The final Tostock charter in the Sibton Abbey Cartulary records that William, the son of Roger of Tostock, received from Abbot Alexander of Sibton the monastery's mill in Tostock with its site and appurtenances. They were granted as a fief to be held for twelve shillings a year, and William and his heirs agreed to maintain them in good condition forever.[7] Significant changes had occurred since the previous charters were composed. The mill was no longer specifically termed a windmill—a reversion to looser linguistic usage. William, the son of Wilmer, was not mentioned. The annual rental had soared, presumably as a result of inflation. And William the son of Roger of Tostock was evidently quite a different person from the original wind-power pioneer, William the son of Roger of Norton. That Roger of Norton may well have been the R—— of Norton who witnessed the grant of Timworth windmill to the Abbey of Bury Saint Edmunds.[8] Timworth is only a few miles from Tostock.

Abbot Alexander, who rented the Tostock post-mill back to lay individuals rather than have his monks manage it directly, was an elusive figure. He ruled Sibton in 1212, but little else is known about him.[9] His decision to alienate the mill was probably determined by Cistercian regulations, reaffirmed by general chapters of the order in 1214 and 1215, which dictated that when such facilities were not directly used by the monks, they were to be sold or leased to someone else.[10] The Tostock windmill was particularly vulnerable to such policies because it was a long distance from the abbey and from its other farms.

Alexander was a shrewd businessman. The twelve-shilling rent he demanded was sixteen times the nine pence earlier demanded for the Tostock mill. Moreover, the abbot's agreement with William the son of Roger of Tostock contained a clause allowing him reentry to the mill and reimbursement for any breach of contract.

The six Tostock charters span a number of years, but the age of the windmill is still indefinite. It is also regrettable that the documents do not reveal more about the trials of Robert Danmartin's wife.

1. B.L. Arundel MS. 221 (Sibton Abbey Cartulary), fol. 41v.

2. Ibid., fol. 42.

3. The details about Robert Danmartin were traced by Philippa Brown, editor of *Sibton Abbey Cartularies and Charters*, 1:48, who is currently editing the Sibton documents. The charters that I cite from the Arundel manuscript will evidently be charters 70–75 in her future volumes. On Eustace the monk, see *Pipe Rolls, 3 Henry III*, p. 47 (Michaelmas 1219); and Henry Lewis Cannon, "The Battle of Sandwich and Eustace the Monk." Before 1154 King Stephen employed a clerk named Robert of Boulogne. He held a canonry in the collegiate church of Saint Martin-le-Grand, London, as had his father before him; *Regesta*, vol. 3, charters 536, 556. He seems a bit early to be identified with Robert the Marshal of Boulogne, but one can never be sure.

4. B.L. Arundel MS. 221, fol. 42.

5. Ibid., fol. 42v.

6. Ibid.

7. Ibid.

8. See Appendix entry no. 47.

9. Knowles, Brooke, and London, *Heads of Religious Houses*, p. 142; *Sibton Abbey Cartularies*, 1:4. A successor is known to have been appointed by 1228.

10. *Sibton Abbey Cartularies*, 1:125, 134.

## 49. SUFFOLK: WILLINGHAM   −1200

In 1202 a dispute between two brothers over inheritance of the land of their late father, Robert de Munti, was settled in the royal court in

Suffolk. By the terms of the agreement, Ralph de Munti, undoubtedly the older brother, was recognized as having the legitimate claim to all of the father's possessions, including one knight's fee in Willingham.

To obtain the settlement, Ralph gave to his brother, Waleran, one-third of that fee, which was his for his lifetime, provided the rights of their mother, Emma, were protected. Included in Waleran's third was a windmill (*unum molendinum ventile*), located on one acre between Hup-chirche and the croft, or garden, belonging to Hugh Rigolf. Waleran was not to sell or give away the property, and at his death it was to revert to Ralph, or to his heirs. If Ralph were to die first, and without heirs, Waleran would inherit everything, except the rights of Ralph's wife.[1]

It is clear from the amount of time such cases usually required to reach the court that the post-mill existed at least a few years before 1202. Therefore, it is much more likely to have been constructed by Robert de Munti than by either of his sons. The Willingham post-mill remained entirely in lay hands. It was one of the few so recorded, but there were probably many others.

Medieval spelling was very inconsistent. Beatrice, daughter of Ralph de Mumby, once owned the Hogsthorpe windmill in Lincoln-shire. It is conceivable that her father was the Ralph de Munti who was a party to this settlement.[2]

1. *Feet of Fines, Suffolk*, p. 161, item no. 330.
2. See Appendix entry no. 17.

## 50. SURREY: WARLINGHAM   −1200

Odo of Danmartin, the son of William of Danmartin, founded the Hospital of Saint James at Tandridge, evidently before the end of the twelfth century. This institution, which was under the direction of the Augustinian canons, was intended for the poor and the sick, and for travelers. His gift to the hospital included all his lands in War-lingham, as well as a windmill (*cum molendino ad ventum*) with its ap-purtenances, free from all service he owed to the heirs of William of Hammes. Although Odo's charter was attested by a dozen witnesses, I cannot date it precisely. William Dugdale's belief that it was executed during King Richard's reign (1189–1199) seems tenable.[1]

Odo's family were vigorously engaged in charitable work. His fa-ther evidently supported Merton Priory and, before 1209, his daughter endowed the Tandridge hospital. The family held land in Essex, Norfolk, and Suffolk, as well as Surrey, and Robert Danmartin once owned the Tostock windmill in Suffolk.[2] Odo was mentioned as the founder of Tandridge hospital in a charter of W(alter), prior of Merton Priory (1198–1218), apparently datable to June 1217.[3] During the tenure of Thomas de Welst as prior of Merton (1218–1222), Odo gave the hospital some silver cups, books, vestments, and cattle.[4] Shortly thereafter, presumably following Odo's death, Saint James Hospital, like many other small Augustinian hospitals, abandoned its medical responsibilities and became an Augustinian priory.

A charter dated 1218 records the gift to Merton Priory of a windmill (*molendinum nostrum in Perang per ventum*) by William of Aguillon.[5] This gift was confirmed by Prior Thomas.[6] The post-mill was probably much older, however, because William was active long before 1218[7] and because he reported that a Bartholomew of Kensington had previously rented the mill for two shillings.

1. Dugdale, *Monasticon Anglicanum*, 6:603–4; Alfred Heales, *Tandridge Priory and the Austin Canons*, p. 23.

2. *V.C.H.*, *Surrey*, 2:94. On the Tostock mill, see Appendix entry no. 48.

3. *V.C.H.*, *Surrey*, 2:112. The charter is in B.L. Cotton MS. Cleopatra C VII (Merton Priory Cartulary), fol. 86. The date is written in a later hand. If the prior were W(illiam), the date would be 1167–1177.

4. Dugdale, *Monasticon Anglicanum*, 6:606.

5. B.L. Cotton MS. Cleopatra C VII, fol. 86v.

6. Ibid.

7. For example, the name of William Aguillon appears in *Receipt Roll of the Exchequer for Michaelmas Term 1185*, p. 4.

## 51. SUSSEX: AMBERLEY    1180–1185

Between December 1184 and December 1185, Bishop Seffrid II of Chichester gave to the canons of his cathedral a windmill (*molendinum venti*) that he had "first caused to be built" at Amberley in Sussex.[1] Since Seffrid was consecrated a bishop in 1180, the dating of this post-mill is unusually precise.

1. Henry Mayr-Harting, *The Acts of the Bishops of Chichester, 1075–1207*, p. 145, document no. 85; *Chichester Cartulary*, p. 41, charter 177. On Seffrid's career, see Henry Mayr-Harting, *The Bishops of Chichester, 1075–1207*, pp. 14–16. For more information about this windmill, see Chapter 8.

## 52. SUSSEX: BOXGROVE    c. 1180

Boxgrove Priory, the Benedictine monastery founded by Robert de Haye in 1105, was not far from Chichester Cathedral. A windmill near the house was not identified as such until about 1225.[1] However, about 1180 Roger Hay had given the monks free access to their mill across enclosed fields.[2]

Boxgrove Priory also owned a windmill at Avisford, donated before 1212–1222, and at least two others at East Marden, donated before 1244, but it is not known how much earlier than these dates they were constructed.[3]

1. *Boxgrove Cartulary*, p. 67, charter 91 (during Prior Ansketill's tenure, c. 1216–1252). The grant was attested by men who were well known in the mid-twelfth century. The cartulary editor suggested that the date was c. 1225. He also was convinced that the windmill existed before 1180; see his entries in the Index under "Boxgrove."

2. Ibid., p. 83, charter 137. About 1180, Roger also gave the monks a water mill at Hunston and allowed them free access to that mill; p. 28, charter 21; p. 30, charter 26; p. 62, charter 77; p. 84, charters 140–41; p. 86, charter 146. Nicholas of Hunston gave the monks a way to the Boxgrove mill (p. 90, charter 160), but his charter does not seem datable.

3. Ibid., p. 77, charter 122; p. 147, charter 328. The Avisford mill was probably donated considerably earlier than the other mills because witnesses to its grant are known to have lived about 1180. Several of the priory's charters concerning its various mills, including the Boxgrove and Avisford windmills, were attested by members of the Aguillon family. William of Aguillon gave the canons of Merton Priory a windmill; see Appendix entry no. 50.

## 53. SUSSEX: ECCLESDON    –1183

The records pertaining to the post-mill at Ecclesdon in Ferring (*Feringis*) are unusual. They yield information about the mill in the planning stage, the specific individuals who constructed it, its finished appearance, and parts of its history.

Sometime before 1183 the dean and chapter of Chichester Cathedral contracted with William de Vescy to build a mill on their land in Ecclesdon. He was to pay a yearly rental of one mark, but he died in 1183 before the project was completed.[1] Bishop Seffrid II resolved to undertake the work himself and was delighted with its results. Sometime before 1197 he gave the windmill (*molendinum ad ventum*) to his steward, Thomas of Ferring, who was asked to pay the cathedral canons a pound of pepper a year.[2] In the charter, Seffrid boasted that he had constructed this windmill at his own expense and that he was including two acres of land next to the road that belongs to the mill and the breadth of one acre around "the outer end of the beam by which the mill is turned round." He also allowed free access to his land from every part of the down for those who wished to use the mill.

Thomas subsequently gave the mill to the Augustinian priory of Tortington. Sometime before 1197, those canons sold the post-mill back to the cathedral chapter for twenty-five marks.[3]

Bishop Seffrid also gave the chapter a water mill at Bishopstone.[4] A successor, Bishop Ranulf (1217–1222), built a windmill there and donated it to the chapter. The wording of the charter almost repeated Seffrid's. Ranulf declared that he was giving the chapter the site of the windmill together with "the breadth of a rod of land outside the outer end of the beam by which the mill is turned round."[5] Note that he gave rods of land, whereas Seffrid had given acres. In later centuries chroniclers would confuse these donations and erroneously claim that Bishop Seffrid had given the chapter the windmill in Bishopstone.[6]

1. *Chichester Cartulary*, p. 95, charter 362. See also Chapter 8.

2. Ibid., p. 42, charter 180; Mayr-Harting, *Acts of the Bishops*, p. 147, document no. 87.

3. *Chichester Cartulary*, p. 43, charter 182. The sale was transacted during the tenures of Prior Nicholas of Tortington (1180–1197) and Dean Matthew of Chichester.

4. Mayr-Harting, *Acts of the Bishops*, p. 146, document no. 86 (1189–1197); *Chichester Cartulary*, p. 42, charter 179 (1189–1191). The mill had belonged to Pain the clerk, then to Reginald of London, and then to Seffrid, who paid five marks for it. Also see *Chichester Cartulary*, p. 43, charter 181 (1189–1197). On Reginald of London, see *St. Bartholomew's Cartulary*, p. 149, charter 1367 (c. 1200).

5. *Chichester Cartulary*, p. 50, charter 206 (1218–1222). The mill's site has not been clearly identified; see Martin Bell, *Excavations at Bishopstone*. For other references to the windmill, see *Chichester Cartulary*, p. 93, charter 352 (a claim to a windmill made by one of Reginald of London's heirs); p. 94, charter 353 (the sale of both mills back to the chapter for sixteen marks). In 1243 (or 1293—the record is unclear), the canons' profits from both the windmill and the water mill totaled eighty shillings; p. 119, charter 458.

6. *Chichester Cartulary*, p. 277, charter 903 (includes a catalog dated c. 1375).

## 54. WARWICKSHIRE: WIBTOFT   −1189

In order to ensure his burial in Leicester Abbey, Ralph Arraby gave the Augustinian canons there one-third of the village of Wibtoft. Included were two mills, a windmill, and a horse mill (*cum duobus molen-*

*dinis uno ad ventum et alio ad equos*).[1] The grant was said to be listed in one of Henry II's confirmations; thus it was made before 1189.[2] Arnold Wood, the steward of the earl of Leicester, confirmed Ralph's grant.[3] Evidently he held property in Wibtoft.

Wibtoft is directly across the road from Wigston Parva, Leicestershire, the location of the earliest known post-mill. Ralph Arraby managed that windmill sometime after 1154. Arnold had held the manor of Wigston Parva before 1154.[4]

There may have been a second early windmill in Wibtoft, for William Pykard gave Leicester Abbey some land and the windmill on the hill there (*molendinum ventricum super montem ibidem*). This grant seems to have been made after 1200.[5]

1. A few other lands were also included. A fifteenth-century Leicester Abbey rental survey includes two references to this grant, but not a copy of the relevant charter; Oxford University, Bodleian Library Laud MS. Misc. 625 (Leicester Abbey Rental), fols. 8, 144. According to another Leicester register, B.L. Cotton MS. Vitellius F XVII (Leicester Abbey Rental), fol. 25, Ralph Arraby gave the canons Wibtoft and a mill (unspecified) there. On fol. 41v, it is stated that Ralph Arraby gave all his land in Wibtoft and Knighton (*Knyton*), but no mention is made of a mill. John Nichols, in *The History and Antiquities of the County of Leicester*, 4:123, reported that Ralph Arraby gave one-third of Wibtoft, some other tenements, and a mill. The grant is similarly described in *V.C.H. Warwickshire*, 6:258–59. Nichols (*History and Antiquities*, vol. 1, pt. 2, appendix, p. 71) also said Ralph Arraby granted one-third of Wibtoft as well as one virgate in Stretton held by Roger Torr. I have not found the grants Nichols described in either register.

2. This notice was added to the entry on fol. 144 of Bodleian Library Laud MS. Misc. 625, but I have not found the royal charter. A long confirmation of Leicester Abbey privileges granted by Henry II, issued about 1168, was copied on fols. 6v–7, but Wibtoft was not mentioned. Presumably, it therefore was acquired by the abbey between 1168 and 1189.

3. William Famcot also confirmed Ralph's grant; ibid., fol. 144. Ralph attested one of Arnold's writs about Claybrook Church; B.L. Additional Charter 47513 T (dated about 1160).

4. On Wigston Parva, see Chapter 3 and Appendix entry no. 15.

5. Oxford University, Bodleian Library Laud MS. Misc. 625, fol. 144.

## 55. YORKSHIRE: BEEFORD  c. 1170

In the mid-twelfth century, Holderness peninsula on the North Sea was a low-lying area that was slowly being drained by numerous dikes. The feudal overlord, the count of Aumale, parceled out the land among his retainers. Geoffrey Brito received two carucates (about 240 acres) near Beeford. He eventually provided for his four daughters by dividing the land equally among them. The eldest girl, Emelina, married Roger of Greynesbury and they had a son, Thomas.

A Cistercian abbey was established at nearby Meaux in 1151 as a partial compensation for the inability of its founder, the overweight Count William le Gros (1130–1179), to go on the Second Crusade, as he had vowed. In time, Roger and his wife sold their acres to the new community for thirteen marks. However, they thriftily retained for their own use two mills—one windmill and one water mill (*duo molendina, unum ventricium at aliud aquaticum*)—and the sluice. These mills were later sold to an unnamed count of Aumale. He gave the mills to the abbey just before he left for Jerusalem. The Meaux community gradually acquired the other parts of Geoffrey Brito's original tenement and converted the land into a grange for the abbey.

There were three counts of Aumale named William in the twelfth century, and as so frequently happens, the chronology of events pertaining to the two mills became confused over the years. More than two hundred years after these transactions, Abbot Thomas Burton (1396–1399) detailed them in his house chronicle, using the elaborate terminology just mentioned. Thomas gave two different accounts. According to one, the sales and gifts were made during the tenure of Abbot Philip (1160–1182); according to the other, they were made under Abbot Thomas (1182–1197).[1]

In the first version, Thomas maintained that sometime after they had received the half carucate Roger, his wife, and their son sold it (but not the mills) to the monks and surrendered any further rights to the land to a Count William of Aumale. Peter, the son and heir of Thomas of Greynesbury, later sold the two mills and sluice to the count. That lord kept them for a while and then gave them to the abbey before setting out for the Holy Land.

In the second version, Thomas stated that Roger of Greynesbury sold the land in the time of Count William de Forz (1190–1195) and that Roger's son, Thomas, later sold the windmill and the water mill to the earl, who gave them to Meaux Abbey in free alms.

One conclusion can be drawn from these confusing reports: the windmill had been transferred to Meaux Abbey at least by 1191, when Count William de Forz left for the Third Crusade. Considering that the mill had several probable owners (the earl, Peter the grandson, Thomas the son, Roger of Greynesbury and his wife, and perhaps Geoffrey

Brito), by 1191 the post-mill must have been in existence for many years.

The two versions suggest that Roger and Emelina sold their land to Meaux Abbey during Philip's tenure (1160–1182), but that the windmill did not arrive at the abbey until 1190–1191, early in the tenure of Abbot Thomas. Therefore, the post-mill must have been in existence in 1182. Determination of Roger of Greyncsbury's date of death might permit verification of an earlier date. Accordingly, it seems reasonable to conclude that the post-mill had been constructed by at least 1170.

Assuming that the windmill had periodic repairs, it probably operated at least two hundred years. About 1380, the monastery moved the windmill to another site to avoid continued payment of a tithe on its profits to the rector of Beeford.[2]

Meaux Abbey owned a number of other windmills. Between 1249 and 1269, one such mill was located beside the abbey's water mill at Ashdyke-on-the-Hull. The monks planned to have a miller and his young male assistant operate both facilities, but the watercourse kept overflowing.[3] By 1260, another post-mill was located at Ravenser Odd, on land reclaimed from the sea. Two others existed at Myton before 1293.[4]

1. Thomas of Burton, *Chronicle*, 1:164–65 (on Abbot Philip); also pp. 224–25 (on Abbot Thomas).

2. Ibid., 3.1/2, 185.

3. Ibid., 2:82. Bennett and Elton, in *History of Cornmilling*, 2:241; and Barbara English, in *The Lords of Holderness, 1086–1260*, p. 207, both mentioned this windmill.

4. English, *Lords of Holderness*, p. 213; Maurice Beresford, *New Towns of the Middle Ages*, p. 516.

## 56. YORKSHIRE: WEEDLEY   −1185 (perhaps −1155)

In 1185 the Knights Templar cataloged many of the estates and other gifts they had received since the order's establishment in England in 1128. Among their numerous possessions was a post-mill (*molendinum venti*) at Weedley (*apud Withele*), in the East Riding of Yorkshire. The mill was at South Cave near the Humber River, not far from what is today Kingston upon Hull. The windmill was then leased by Nicholas, son of Alexander of Hibaldstow, for an annual rent of eight shillings.[1] Many writers have claimed that this mill is the oldest identifiable wind-powered facility. This book amply demonstrates that such a conclusion is no longer valid.

Alexander came from Hibaldstow, in Lincolnshire. A tenant named Gamel held a water mill from him in Drewton, North Cave, not far from Weedley.[2] A windmill was located in Drewton at least by 1212.[3] A few other post-mills may have existed in the immediate vicinity.[4]

The Weedley windmill was definitely constructed before 1185—possibly thirty years earlier. Roger de Mowbray and his mother were among the Templars' most enthusiastic supporters. Roger founded three preceptories. Among his smaller benefactions was a donation of carucates (about 360 acres) in Weedley, made before 1154.[5] His grants were recorded in the Templar survey without any indication of a date. Between 1154 and 1157, to compensate York Minster for the depradations it had suffered during the recent civil war, Roger donated to the cathedral the tithes on the profits of all the mills in South Cave.[6] This proves that the Templars held a large proportion of Weedley for approximately thirty years, that mills had existed in South Cave for the same length of time before the inventory, and that the Weedley windmill might have been one of those mills.

Weedley is now a deserted medieval village that has reverted to plowland. Each year its newly tilled soil reveals a circular scatter of white that may be evidence of a buried windmill substructure.[7] This hamlet has not yet been excavated, but if it is, more precise dating of the Weedley windmill may be possible.

1. Lees, *Records of the Templars*, p. 131; also printed in Dugdale, *Monasticon Anglicanum*, 6:821. For the original transcript, written in a clear hand, see P.R.O. MS. E 164.16 (formerly called Exchequer K. R. Miscellaneous Books 16).

2. Lees, *Records of the Templars*, p. 131.

3. *Curia Regis Rolls*, 6:373; also see *V.C.H., Yorkshire, East Riding*, 4:32.

4. Dugdale, *Monasticon Anglicanum*, 6:831, erroneously printed that at Baggaflet (probably Broomflete, South Cave) Richard (apparently another son of Alexander of Hibaldstow) leased a *molendinum venti* for two marks. The original manuscript simply read "a mill" and Lees edited that.

5. Lees, *Records of the Templars*, pp. 125–26; Diana E. Greenway, ed., *Charters of the Honour of Mowbray, 1107–1191*, charter 273; Farrar and Clay, *Early Yorkshire Charters*, vol. 1, charter 185.

6. Greenway, *Charters of the Honour of Mowbray*, charter 324. This donation was witnessed by a number of men from Cave and by William the physician, a canon of London. See also Farrar and Clay, *Early Yorkshire Charters*, 3:435, charter 1825.

7. Maurice Beresford and John G. Hurst, *Deserted Medieval Villages*, pp. 180–81.

# *Abbreviations*

| | |
|---|---|
| B.L. | British Library |
| C.U.L. | Cambridge University Library |
| D.L. | Duchy of Lancaster |
| *EcHR* | *Economic History Review* |
| *EHR* | *English Historical Review* |
| *JEH* | *Journal of Ecclesiastical History* |
| H.M.C. | Historical Manuscripts Commission |
| P.R.O. | Public Record Office |
| *V.C.H.* | *The Victoria History of the Counties of England* |
| *TNS* | *Transactions of the Newcomen Society for the Study of the History of Engineering and Technology* |

# Bibliography

Numerous sources were investigated for this study, but the following provided the most pertinent information. Unedited medieval manuscripts are classified by their current repositories. Published chronicles are cited with the names of original authors whenever possible. Anonymous cartularies are alphabetized by the customary names of their religious houses (e.g., *Clerkenwell Cartulary*). Twelfth-century government records are filed under their working titles (e.g., *Feet of Fines, Kent*).

## MANUSCRIPTS

Cambridge. Cambridge University, King's College Manuscripts: Bricett Deeds, B 4, B 8, B 10, B 11, B 12, B 13, B 16, B 20, B 69, B 79, B 94, B 95/1, B 95/2, B 105, B 114 group 1 and group 5.

Cambridge University Library Manuscripts: 3021 (Thorney Abbey Cartulary), Dd. 3.87 (Owston Abbey Cartulary), Ee. 2.31 (Aristotle's works on natural philosophy), Ff. 2.33 (Bury Sacristan's Register), Mm. 2.20 (Bromholm Priory Cartulary), Mm. 4.19 (Black Book of Bury), Additional 4220 (Bury Cellarer's Cartulary), Additional 6845 (Saint Lawrence Hospital Register).

Canterbury. Canterbury Cathedral Muniments, Ancient Charters: C 23, C 712, C 1088, C 1161, C 1184, C 1207, E 133, F 42, M 130, M 131.

Register A, Register C, Register E.

Literary Manuscripts: B 14 (Canterbury Rental), B 15 (Canterbury Rental), C 20 (Saint Lawrence Hospital Register), D 4 (Canterbury Rental).

Canterbury. Canterbury Cathedral Archives. Eastbridge Hospital Muniments, Ancient Charters: A 2, A 15, A 26, A 46, A 70, A 71, A 76, B 27, B 59, C 1, C 2, C 3, C 4, C 5, C 23, F 20.

Cyprian Bunce, A Register of Loans and Charitable Donations to the Poor of Canterbury (handwritten calendar, 2 vols. Canterbury, 1798).

London. British Library, Additional Charters: 19571, 19591, 28340, 28348, 45760, 47398, 47513T, 47551, 47552, 47556, 47557, 47558, 47559, 47560, 47573R, 47573T, 47573U, 47580, 47599, 47639, 48086, 48137, 48296.

Cotton Charters: XXI.6, XXV.23.

Harley Charters: 45 I 52, 45 I 53, 45 I 56, 45 I 77, 46 A 7, 55 G 9.

Additional Manuscripts: 4934, 4935 (transcripts of Croxton Abbey Registers), 5516 (Holy Sepulchre Priory Cartulary), 14847 (White Book of Bury), 14848 (William Curtey's Bury Register), 29437 (Poor Priests' Cartulary), 35296 (Spalding Priory Cartulary), 37022 (Pipewell Abbey Cartulary), 42130 (Luttrell Psalter), 46353 (Dereham Abbey Cartulary), 46701 (Stixwould Priory Cartulary), 47682 (*The Holkham Bible Picture Book*).

Arundel Manuscript 221 (Sibton Abbey Cartulary).

Cotton Manuscripts: Julius A VI (Calendar), Tiberius B V (Calendar), Claudius A VI (Boxgrove Priory Cartulary), Claudius D X (Red Book of Saint Augustine's Abbey), Claudius D XII (Daventry Priory Cartulary), Nero C IX (Hospitaller Cartulary), Nero C XII (Walsall Parish Cartulary), Nero E VI (Hospitaller Cartulary), Vitellius F XVII (Leicester Abbey Rental), Vespasian E V (Reading Abbey Cartulary), Vespasian E XVIII (Kirkstead Abbey Cartulary), Vespasian E XX (Bardney Abbey Cartulary), Vespasian E XXV (Reading Abbey Cartulary), Titus C VIII (Wymondham Priory Cartulary), Cleopatra C VII (Merton Priory Cartulary), Cleopatra C XI (Saint Anselm, *De Similitudinibus*), Faustina A IV (Saint Neot's Priory Cartulary), Faustina B I (Barlings Abbey Cartulary), Appendix XXI (Stoke-by-Clare Priory Cartulary).

Egerton Manuscript 3031 (Reading Abbey Cartulary).

Harley Manuscripts: 662 (Dunmow Priory Cartulary), 742 (Spalding Priory Cartulary), 1708 (Reading Abbey Cartulary), 2110 (Castle Acre Priory Cartulary), 3487 (Aristotle's works on natural philosophy), 4748 (Segrave baronial cartulary index).

Lansdowne Manuscripts: 207 E (Kirkstead Abbey Cartulary transcript), 424 (Meaux Abbey Cartulary).

Royal Manuscripts: 2 B VII (Queen Mary's Psalter), 10 E IV (Smithfield Priory Decretals).

Stowe Manuscript 928 (Croxton Abbey Register transcript).

London. Guildhall Museum, Saint Paul's Cathedral Muniments, Ancient Deeds: Press A, Box 2, no. 212; Press A, Box 40, no. 1412.

Manuscripts: W.D. 1 (Liber A, Pilsosus), W.D. 4 (Liber L).

London. Public Record Office, Ancient Charters: Duchy of Lancaster D.L. 25/3073; D.L. 25/3090; D.L. 36/2, charter 22; D.L. 36/2, charter 113.

Ancient Deeds: E 315/43, charter 39; E 315/46, charter 1.

Miscellaneous Charter 249.

Manuscripts: D.L. 42.1 and D.L. 42.2 (Duchy of Lancaster Coucher Book), D.L. 42.5 (Bury Cellarer's Cartulary), E 164.16 (Knights Templar Survey of 1185), E 164.20 (Godstow Abbey Cartulary).

New York City. Pierpont Morgan Library, Manuscript 102 (Windmill Psalter).

Oxford. Oxford University, Bodleian Library, Ancient Charters: Norfolk Charter 424, Suffolk Charter 239.

Manuscript 264 (*The Romance of Alexander*).

Dodsworth Manuscript 135 (miscellaneous transcripts).

Laud Manuscript, Misc. 625 (Leicester Abbey Rental).

Tanner Manuscript 425 (Hickling Priory Cartulary).

Topham Kent Manuscripts: c.2 (Eastbridge Hospital Register), d.3 (Saint Lawrence Hospital Register).

Oxford. Oxford University, Christ Church College Manuscript 92 (*The Treatise of Walter de Milemete*).

Oxford. Oxford University, Magdalene College Muniments, Ancient Charters: Romney Marsh Deeds nos. 15, 29, 30, 31, 37, 41, 42, 46, 53, 57, 59, 60, 61, 62. William Dunn Macray, Typescript Catalog of Romney Marsh Deeds.

## PUBLISHED EDITIONS OF MEDIEVAL SOURCES

Adams, Norma, and Donahue, Charles, Jr. *Select Cases from the Ecclesiastical Courts of the Province of Canterbury, 1200–1301.* Publications of the Selden Society 95 (1981, for 1978–1979).

Ballard, Alphonsus, ed. *British Borough Charters, 1042–1216.* Cambridge, England, 1913.

Barfield, Samuel. "Lord Fingall's Cartulary of Reading Abbey." *EHR* 3 (1888): 113–25.

Barnes, Patricia M., and Powell, W. Raymond, eds. *Interdict Documents.* Pipe Roll Society Publications 72, n.s. 34 (1960 for 1958).

*Battle Chronicle.* Searle, Eleanor, ed. and trans. *The Chronicle of Battle Abbey.* Oxford: Oxford University Press, 1980.

Berger, Elie, ed. *Recueil des actes de Henri II concernant de provinces françaises et les affaires de France.* Posthumously published work of Leopold Delisle. 3 vols. Paris, 1909–1927.

Birch, Walter de Gray. *Cartularium Saxonicum: A Collection of Charters Relating to Anglo-Saxon History, A.D. 430–975.* 4 vols. London, 1885–1899.

Bishop, T. A. M. *Scriptores Regis.* Oxford: Oxford University Press, 1961.

*Blythburgh Cartulary.* Harper-Bill, Christopher, ed. *Blythburgh Priory Cartulary.* 2 vols. Woodbridge, Suffolk: Boydell and Brewer, for the Suffolk Records Society, 1980, 1981.

*Boxgrove Cartulary.* Fleming, Lindsay, ed. *The Chartulary of Boxgrove Priory.* Sussex Record Society Publications 59 (1960).

*Calendar of the Charter Rolls, 1226–1516.* 6 vols. London, 1903–1927.

Cam, Helen C. "An East Anglian Shire-moot of Stephen's Reign, 1148–1154." *EHR* 39 (1924): 568–71.

*Charlemagne, An Anglo-Norman Poem of the Twelfth Century.* Edited by Francisque Michel. London, 1836.

Chibnall, Marjorie. *Charters and Custumals of the Abbey of Holy Trinity, Caen.* Oxford: Oxford University Press for the British Academy, 1982.

*Chichester Cartulary.* Peckham, W. D., ed. *The Chartulary of the High Church of Chichester.* Sussex Record Society Publications 46 (1946).

*Christ Church Cartulary.* Denholm-Young, Noel, ed. *Cartulary of the Medieval Archives of Christ Church.* Oxford, 1931.

*Clerkenwell Cartulary*. Hassall, W. O., ed. *Cartulary of St. Mary Clerkenwell*. Royal Historical Society Publications, 3d ser. 71 (1949).

*Colchester Cartulary*. Moore, Stuart A., ed. *Cartularium Monasterii Sancti Johannis Baptistae de Colecestria*. 2 vols. London: Roxburghe Club, 1897.

*Colne Cartulary*. Fisher, John L., ed. *Cartularium Prioratus de Colne*. Essex Archaeological Society Occasional Papers 1 (1946).

Coxe, H. O., and Turner, W. H., eds. *Calendar of Charters and Rolls Preserved in the Bodleian Library*. Oxford: Oxford University Press, 1878.

*Curia Regis Rolls*. Preserved in the Public Record Office, by regnal years. London, 1922–.

*Decretalium D. Gregorii IX et Privilegi Gregorii XIII*. London, 1606.

Dodwell, Barbara. *The Charters of Norwich Cathedral Priory*. Pipe Roll Society Publications, n.s. 40 (1965).

*Domesday of St. Paul's*. Hale, William H., ed. *Domesday of St. Paul's of the Year 1222*. Camden Society Publications, o.s. 69 (1858).

Douglas, David C. *The Social Structure of Medieval East Anglia*. Oxford: Oxford University Press, 1927.

———. *Feudal Documents from the Abbey of Bury St. Edmunds*. London: For the British Academy, 1932.

Douglas, David C., and Greenaway, George W., eds. *English Historical Documents*. Vol. 2 *(1042–1189)*. London: Eyre and Spottiswoode, 1953.

Dugdale, William. *Monasticon Anglicanum*. Edited by John Caley, Henry Ellis, and Buckley Bandinal. 6 vols. in 8 pts. London, 1817.

*Dunmow Cartulary*. Levy, Richard Edmond, ed. "The Cartulary of Little Dunmow Priory." 2 vols. M.A. thesis, University of Virginia, June 1971.

Elvey, G. R., ed. *Luffield Priory Charters*. 2 vols. Buckinghamshire Record Society 15 (1968); 22 (1975). Also Northamptonshire Record Society 22 (1968); 26 (1975).

*Epistolae Cantuariensis*. In William Stubbs, ed., *Chronicles and Memorials of the Reign of Richard I*. 2 vols. *Rolls Series* 38 (1864–1865).

*Fair Em*. The Tudor Facsimile Texts. Edinburgh and London, 1911. New York: AMS Reprint, 1970.

Fantosme, Jordan. *Chronique de la guerre entre les Anglois et les Ecossais*. In Richard Howlett, ed. *Chronicles of the Reigns of Stephen, Henry II,* and Richard I. 4 vols. Rolls Series 82 (1885–1890), vol. 3, pp. 202–377.

Farrer, William, and Clay, Charles T., eds. Early Yorkshire Charters. 12 vols. Yorkshire Archaeological Society Record Series, 1914–1965.

Feet of Fines. 7–8 Richard I, 1196–1197. Pipe Roll Society Publications 20 (1896).

Feet of Fines. 9 Richard I, 1197–1198. Pipe Roll Society Publications 23 (1898).

Feet of Fines. 10 Richard I, 1198–1199. Pipe Roll Society Publications 24 (1900).

Feet of Fines, Essex. R. E. G. Kirk and E. F. Kirk, eds. Feet of Fines for Essex. Vol. I (1182–1281). Essex Archaeological Society Publications. Colchester, 1899.

Feet of Fines, Kent. Churchill, Irene J.; Griffin, R.; and Hardman, F. W., eds. Calendar of Kent Feet of Fines to the End of Henry III's Reign. Kent Archaeological Society, Kent Records 15 (1939–1940).

Feet of Fines, Norfolk. Walter Rye, ed. Penes Finium Relating to Norfolk, 3 Richard I to the End of the Reign of John. Norfolk and Norwich Archaeological Society Publications. Norwich, 1881.

Feet of Fines, Norfolk. Barbara Dodwell, ed. Feet of Fines for the County of Norfolk, 1198–1202. Pipe Roll Society Publications 65 (1950).

Feet of Fines, Norfolk. Barbara Dodwell, ed. Feet of Fines for the County of Norfolk for the Reign of King John, 1201–1215. For the County of Suffolk for the Reign of King John, 1199–1214. Pipe Roll Society Publications 70 (1956).

Feet of Fines, Northumberland. Arthur M. Oliver and Charles Johnson, eds. Feet of Fines, Northumberland and Durham, 1196–1228. Newcastle upon Tyne Record Commission Publications 10 (1933).

Feet of Fines, Suffolk. Barbara Dodwell, ed. Feet of Fines for the County of Suffolk, 1199–1214. Pipe Roll Society Publications 70 (1956).

Flower, C. T. Introduction to the Curia Regis Rolls, 1199–1230. Publications of the Selden Society 62 (1972 for 1943).

Geoffrey of Monmouth. The History of the Kings of Britain. Edited by Lewis Thorpe. London: Penguin Books, 1966.

Gerald of Wales. The Journey through Wales and the Description of Wales. Translated by Lewis Thorpe. London: Penguin Books, 1978.

Gervase of Canterbury. Opera: The Historical Works of Gervase of Canter-

*bury.* Edited by William Stubbs. 2 vols. *Rolls Series* 73 (1879–1880).

Gibbs, Marion, ed. *Early Charters of the Cathedral Church of St. Paul's, London.* Royal Historical Society Publications, 3d ser. 58 (1939).

*Gilbert of Hoyland.* Bracelond, Lawrence C., ed. *The Works of Gilbert of Hoyland.* Vol. 1, *Sermons on the Song of Songs.* Kalamazoo, Mich.: Cistercian Publications, 1979.

Giraldus Cambrensis. *Opera.* Ed. by J. S. Brewer, J. F. Dimock, G. F. Warner. 8 vols. *Rolls Series* 21 (1861–1891).

*Godstow Cartulary.* Clark, Andrew, ed. *The English Register of Godstow Nunnery near Oxford.* 3 vols. Early English Text Society, o.s. 129 (1905), 130 (1906), 142 (1911).

Gorham, George Cornelius, ed. *The History and Antiquities of Eynesbury and St. Neots.* 2 vols. London, 1824.

Greenway, Diana E., ed. *Charters of the Honour of Mowbray, 1107–1191.* London: The British Academy, 1972.

Guerncs [Garnier] of Ponte-Sainte-Maxence. *Vie de Saint Thomas.* Edited by Emmanuel Walberg. Lund, Sweden, 1922.

*Guisboro Cartulary.* Brown, William, ed. *Cartularium Prioratus de Gyseburne.* 2 vols. Surtees Society Publications 86 (1889), 89 (1894).

Hardy, Thomas Duffus, ed. *Rotuli Chartarum in Turri Londoniensi Asservati.* London: The Record Commission, 1837.

Hervey, Francis, ed. *The Pinchbeck Register.* 2 vols. Brighton, 1925.

*H.M.C. Rutland. The Manuscripts of His Grace, the Duke of Rutland, Preserved at Belvoir Castle.* 4 vols. London, 1888–1905.

Holdsworth, Christopher J., ed. *Rufford Charters.* 2 vols. Thoroton Society Record Series 29 (1972), 30 (1974).

*The Holkham Bible Picture Book.* Edited by W. O. Hassall. London: Dropmore Press, 1954.

Holtzmann, Walther. *Papsturkunden in England.* 3 vols. Berlin and Göttingen, 1930–1952.

Horn, Walter, and Born, Ernest, eds. *The Plan of St. Gall.* 3 vols. Berkeley and Los Angeles: University of California Press, 1979.

*Hospitaller Cartulary.* Gervers, Michael, ed. *The Hospitaller Cartulary in the British Library (Cotton MS Nero E VI).* Toronto: Pontifical Institute of Medieval Studies, 1981.

Innocent III, *Letters. The Letters of Pope Innocent III (1198–1216) Con-*

*cerning England and Wales*. Edited by Christopher R. Cheney and Mary G. Cheney. Oxford: Clarendon Press, 1967.

Jacobs, Joseph, ed. *The Jews of Angevin England: Documents and Records from Latin and Hebrew Sources*. London, 1893.

Jaffé, Philip, ed. *Regesta Pontificum Romanorum ab Condita Ecclesia ad Annum post Christum datum MCXCVIII*. 2d ed. 2 vols. Edited by S. Lowenfeld, F. Kaltenbruner, and P. Ewald. Leipzig, 1885–1888.

James, Montague Rhodes, and Jenkins, Claude. *A Descriptive Catalogue of the Manuscripts in the Library of Lambeth Palace*. 5 pts. Cambridge: Cambridge University Press, 1930–1932.

Jocelin, *Chronicle*. Butler, H. E., ed. and trans. *The Chronicle of Jocelin of Brakelond, Concerning the Acts of Samson, Abbot of the Monastery of Saint Edmund*. London: Thomas Nelson and Sons, 1949.

*Karlamagnus Saga*. Edited and translated by Constance B. Hieatt. 3 vols. Toronto: Pontifical Institute of Medieval Studies, 1975–1980.

*Karls des Grossen Reise nach Jerusalem und Constantinopel*. Edited by Eduard Koschwitz. Leipzig, 1913.

Kauffmann, C. M. *Romanesque Manuscripts, 1066–1190*. London: Harvey Miller, 1975.

Kemble, John M., ed. *Codex Diplomaticus Aevi Saxonici*. 6 vols. London: English Historical Society, 1839–1848.

*Kentish Cartulary*. Cotton, Charles, ed. *A Kentish Cartulary of the Order of St. John of Jerusalem*. Kent Archaeological Society, Kent Records 11 (1930).

*Knights' Cartulary*. Gervers, Michael, ed. *The Cartulary of the Knights of St. John of Jerusalem in England. Secunda Camera, Essex*. Oxford: The British Academy, 1982.

Langtoft, Peter. *Chronicle in French Verse*. Edited and translated by Thomas Wright. 2 vols. *Rolls Series* 47 (1866–1868).

Larking, Lambert B., and Kemble, John Mitchell, eds. *The Knights Hospitallers in England*. Camden Society Publications 65 (1857).

Lees, Beatrice A., ed. *Records of the Templars in the Twelfth Century: The Inquest of 1185*. London: The British Academy, 1935.

*Leges Henrici Primi*. Edited and translated by L. J. Downer. Oxford: Oxford University Press, 1972.

*Leiston Cartulary*. Mortimer, Richard, ed. *Leiston Abbey Cartulary and Butley Priory Charters*. Woodbridge, Suffolk: Boydell and Brewer, for the Suffolk Records Society, 1979.

*Liber Eliensis.* Edited by E. O. Blake. Royal Historical Society Publications, 3d ser. 92 (1962).

*Liber Vitae Dunelmensis; nec non obituaria duo eiusdem ecclesiae.* Edited by Joseph Stevenson. Surtees Society Publications 13 (1841). For a facsimile, see *Liber,* ed. A. H. Thompson. Surtees Society Publications 136 (1923).

Loyd, Lewis C., and Stenton, Doris M., eds. *Sir Christopher Hatton's Book of Seals.* Oxford, 1950.

Macray, William D. *Notes from the Muniments of Magdalene College.* Oxford, 1882.

Major, Kathleen, ed. *Acta Stephani Langton Cantuariensis Archiepiscopi, A.D. 1207–1229.* Canterbury and York Society Publications 1 (1950).

Mayr-Harting, Henry. *The Acts of the Bishops of Chichester, 1075–1207.* Canterbury and York Society Publications 130 (1964).

Meekings, C. A. F. "Two Danmartin Deeds." *Sussex Archaeological Society Collections* 116 (1978):399–400.

Migne, J. P., ed. *Patrologiae cursus completus: Series Latina.* 221 vols. Paris, 1844–1864.

*Memoranda Roll, 1 John (1199–1200).* Edited by H. G. Richardson. Pipe Roll Society Publications 59 (1943).

Millar, E. G. *The Luttrell Psalter.* London, 1932.

Morey, Adrian, and Brooke, Christopher N. L. *The Letters and Charters of Gilbert Foliot.* Cambridge: Cambridge University Press, 1967.

Morgan, Nigel. *Early Gothic Manuscripts, 1190–1250.* London: Harvey Miller, 1982.

*Newnham Cartulary.* Godber, Joyce, ed. *The Cartulary of Newnham Priory.* 2 vols. Publications of the Bedfordshire Historical Record Society 43 (1963, 1964).

Oliver, Arthur M., ed. *Early Deeds Relating to Newcastle-upon-Tyne.* Surtees Society Publications 137 (1924).

*Oseney Cartulary.* Salter, H. E., ed. *The Cartulary of Oseney Abbey.* 6 vols. Oxford Historical Society Publications 89 (1929), 90 (1930), 91 (1931), 97 (1934), 98 (1935), 101 (1936).

Owen, Dorothy M., ed. *A Catalog of Lambeth Manuscripts, 889 to 901.* London: Lambeth Palace Library, 1968.

*Pipe Rolls.* Cited by regnal years. London: Pipe Roll Society Publications, 1884–.

Powicke, F. M., and Cheney, Christopher R., eds. *Councils and Synods with Other Documents Relating to the English Church.* Vol. 2 (*1205–1313*). Oxford: Clarendon Press, 1964.

*Ramsey Cartulary.* Hart, William Henry, and Lyons, Ponsonby A., eds. *Cartularium Monasterii de Rameseia.* 3 vols. *Rolls Series* 79 (1884–1894).

*Ramsey Chronicle.* Macray, W. Dunn, ed. *Chronicon Abbatiae Rameseiensis. Rolls Series* 83 (1886).

*Receipt Roll of the Exchequer for Michaelmas Term 1185.* Edited by Hubert Hall. London, 1899.

*Red Book of the Exchequer.* Edited by Hubert Hall. 3 vols. *Rolls Series* 99 (1896).

*Regesta. Regesta Regum Anglo-Normannorum.* Vol. 1 (*1066–1100*), ed. H. W. C. Davis (Oxford, 1913); vol. 2 (*1100–1135*), ed. Charles Johnson and H. A. Cronne (Oxford, 1956); vol. 3 (*1135–1154*), ed. H. A. Cronne and R. H. C. Davis (Oxford, 1968); vol. 4 (facsimiles), ed. H. A. Cronne and R. H. C. Davis. Oxford: Clarendon Press, 1969.

*Register of St. Augustine's Abbey, Canterbury, Commonly Called the Black Book.* Edited by George J. Turner and Herbert E. Salter. 2 pts. London: The British Academy, 1915, 1924.

*Registrum Antiquissimum.* Foster, Charles Wilmer, and Major, Kathleen, eds. *The Registrum Antiquissimum of the Cathedral Church of Lincoln.* 9 vols. Lincoln Record Society Publications 27 (1931), 28 (1933), 29 (1935), 32 (1937), 34 (1940), 41 (1950), 46 (1953), 51 (1958), 62 (1968).

Richard Fitz Nigel. Johnson, Charles, ed. and trans. *The Course of the Exchequer by Richard Son of Nigel.* London: Thomas Nelson and Sons, 1950.

Robertson, James C., and Shepherd, J. B., eds. *Materials for the History of Thomas Becket.* 7 vols. *Rolls Series* 67 (1875–1885).

*Rolls Series. Rerum Britannicarum Medii Aevi Scriptores.* 99 vols. in 253 pts. London, 1858–1896.

Rothwell, Harry, ed. *English Historical Documents.* Vol. 3 (*1189–1327*). New York: Oxford University Press, 1975.

*Rotuli de Dominabus et Pueris et Puellis de XII Comitatibus (1185).* Edited by John Horace Round. Pipe Roll Society Publications 35 (1913).

Round, John Horace. "Bernard, the King's Scribe." *EHR* 14 (1899): 417–30.

————. "A Woodham Ferrers Charter." *Transactions of the Essex Archaeological Society*, n.s. 10 (1909):303–7.

*Saint Bartholomew's Cartulary.* Kerling, Nellie J. M., ed. *The Cartulary of St. Bartholomew's Hospital.* London, 1973.

*Saint Denys Cartulary.* Blake, E. O., ed. *The Cartulary of the Priory of St. Denys near Southampton.* 2 vols. Southampton Records Series 24 (1981).

*Saint Gregory's Cartulary.* Woodcock, Audrey M., ed. *The Cartulary of the Priory of St. Gregory, Canterbury.* Camden Society Publications, 3d ser. 88 (1956).

Saltman, Avrom. *Theobald, Archbishop of Canterbury.* London, 1956.

Samson, *Kalendar.* Davis, R. H. C., ed. *The Kalendar of Abbot Samson of Bury St. Edmunds and Related Documents.* Royal Historical Society Publications, 3d ser. 84 (1954).

*Sibton Abbey Cartularies and Charters.* Vol. 1. Edited by Philippa Brown. Suffolk Records Society, Suffolk Charters 7 (1985).

Smith, D. M., ed. *English Episcopal Acta.* Vol. 1, *Lincoln Diocese (pt. 1, 1067–1185).* Oxford: The British Academy, 1979.

Southern, R. W., and Schmitt, F. S., eds. *Memorials of Saint Anselm.* London: Oxford University Press, for The British Academy, 1969.

Stenton, Doris Mary, ed. *Pleas before the King and His Justices, 1198–1212.* 4 vols. Publications of the Selden Society 67 (1953 for 1948), 68 (1952 for 1949), 83 (1967 for 1966), 84 (1967).

Stenton, Frank M. *The Free Peasantry of the Northern Danelaw.* Oxford: Clarendon Press, 1925, 1969 (reprint).

————, ed. *Documents Illustrative of the Social and Economic History of the Danelaw.* London: The British Academy, 1920.

————, ed. *Transcripts of Charters Relating to the Gilbertine Houses of Sixle, Ormesby, Catley, Bullington, and Alvingham.* Lincoln Record Society Publications 18 (1922).

*Stoke-by-Clare Cartulary.* Harper-Bill, Christopher, and Mortimer, Richard, eds. *Stoke-by-Clare Priory Cartulary.* 2 vols. Woodbridge: Boydell and Brewer, for the Suffolk Records Society, 1982, 1983.

Stow, John. *A Survey of the Cities of London and Westminster and the Borough of Southwark.* Reprinted from the 1603 edition with an introduction by C. L. Kingsford. 2 vols. Oxford, 1908.

Thomas of Burton, *Chronicle.* Bond, Edward A., ed. *Chronica Monasterii de Melsa ab anno 1150 usque ad annum 1506.* 3 vols. *Rolls Series* 43 (1866–1868).

Thomas of Elmham, *Historia*. Hardwick, Charles, ed. *Historia Monasterii S. Augustini Cantuariensis, by Thomas of Elmham, Formerly Monk and Treasurer of that Foundation*. Rolls Series 8 (1858).

Thompson, Alexander Hamilton, ed. *A Calendar of Charters and Other Documents Belonging to the Hospital of William Wyggeston at Leicester*. Leicester, 1933.

Thomson, R. M., ed. *The Chronicle and Election of Hugh, Abbot of Bury St. Edmunds and Later Bishop of Ely*. Oxford: Oxford University Press, 1974.

————. "Twelfth-Century Documents from Bury St. Edmunds." *EHR* 92 (1977):806–19.

Thorne, William, *Chronicle*. *William Thorne's Chronicle of St. Augustine's Abbey, Canterbury*. Translated by A. H. Davis. Oxford, 1934.

*The Treatise of Walter de Milemete: De Nobilitatibus, Sapientiis, et Prudentiis Regum*. Edited by Montague Rhodes James. Oxford: Oxford University Press, for the Roxburghe Club, 1913.

Turner, W. H., and Coxe, H. O., eds. *Calendar of Charters and Rolls Preserved in the Bodleian Library*. Oxford, 1878.

*Walter of Henley and Other Treatises on Estate Management and Accounting*. Edited by Dorothea Oschinsky. Oxford: Clarendon Press, 1971.

*Wardon Cartulary*. Fowler, G. H., ed. *Cartulary of the Abbey of Old Wardon*. Bedfordshire Historical Record Society Publications 13 (1930).

Warner, George, ed. *The Guthlac Roll*. Oxford: Roxburghe Club, 1928.

Warner, G. F., and Ellis, H. J., eds. *Facsimiles of Royal and Other Charters in the British Museum*. Vol. 1 *(William I–Richard I)*. London, 1903.

West, J. R., ed. *The Register of the Abbey of St. Benet of Holme, 1020–1210*. 2 vols. Norfolk Record Society Publications 2, 3 (1932).

William of Malmesbury, *G.R. Willelmi Malmesbiriensis Monachi de Gestis Regum Anglorum*. 2 vols. *Rolls Series* 90 (1887–1889).

*Winton Domesday*. Barlow, Frank; Biddle, Martin; von Feilitzen, Olaf; and Keene, D. J., eds. *Winchester in the Early Middle Ages*. Winchester Studies, vol. 1. Oxford: Clarendon Press, 1976.

Woodruff, C. Eveleigh, ed. "The Register and Chartulary of the Hospital of St. Lawrence, Canterbury." *Archaeologia Cantiana* 50 (1938):33–49.

## MODERN STUDIES

*About Charing: Town and Parish*. By the members of the Charing and District Local History Society for the Kent County Library, 1984.

Acland, James H. *Medieval Structure: The Gothic West*. Toronto: University of Toronto Press, 1972.

Adler, Alfred. "The *Pèlerinage de Charlemagne* in New Light on St. Denis." *Speculum* 22 (1947):550–61.

Alexander, James W. *Ranulf of Chester, A Relic of the Conquest*. Athens: University of Georgia Press, 1983.

Alexander, Jonathan J. *The Decorated Letter*. New York: George Braziller, 1978.

Altschul, Michael. *A Baronial Family in Medieval England: The Clares, 1217–1314*. Baltimore: The Johns Hopkins University Press, 1965.

Baddeley, Welbore St. Clair. "Fresh Material Relating to Norman Gloucester." *Transactions of the Bristol and Gloucestershire Archaeological Society* 41 (1918–1919):86–92.

Barlow, Frank. *The English Church, 1066–1154*. New York: Longman, 1979.

———. *William Rufus*. Berkeley and Los Angeles: University of California Press, 1983.

Batten, M. I., and Smith, Donald. *English Windmills*. 2 vols. London: Architectural Press, 1930–1932.

Bell, Martin. "Excavations at Bishopstone." *Sussex Archaeological Society Collections* 115 (1977):244–50.

Bennett, Adelaide. "The Windmill Psalter: The Historiated Letter E of Psalm One." *Journal of the Warburg and Courtauld Institutes* 43 (1980):52–67.

Bennett, Richard, and Elton, John. *History of Cornmilling*. Vol. 2, *Watermills and Windmills*. London, 1899.

Beresford, Maurice. *New Towns of the Middle Ages*. London, 1967.

Beresford, Maurice, and Hurst, John G. *Deserted Medieval Villages*. New York: St. Martin's Press, 1971.

Beresford, Maurice, and St. Joseph, John K. *Medieval England: An Aerial Survey*. Cambridge: At the University Press, 1958.

Beveridge, W. H. "The Yield and Price of Corn in the Middle Ages." *EcHR* 1 (1926–1929):162–66.

Birch, Walter de Gray. *Catalog of Seals in the Department of Manuscripts of the British Museum.* Vol. 2. London, 1892.

Bloch, Marc. *Land and Work in Medieval Europe.* Translated by J. E. Anderson. Berkeley and Los Angeles: University of California Press, 1967.

Boutflower, D. S. *Fasti Dunelmenses: A Record of the Beneficed Clergy of the Diocese of Durham Down to the Dissolution of the Monastic and Collegiate Churches.* Surtees Society Publications 129 (1926).

Bolton, J. L. *The Medieval English Economy, 1150–1500.* London: J. M. Dent, 1980.

Bott, D. J. "The Murder of St. Wistan." *Transactions of the Leicestershire Archaeological Society* 29 (1953):30–41.

Brown, R. J. *Windmills of England.* London: Robert Hale, 1976.

Butcher, A. F. "The Hospital of St. Stephen and St. Thomas, New Romney: The Documentary Evidence." *Archaeologia Cantiana* 96 (1980):17–26.

Camille, Michael. "Illustrations in Harley MS. 3487 and the Perception of Aristotle's *Libri Naturales* in Thirteenth-Century England." In W. M. Ormrod, ed., *England in the Thirteenth Century: Proceedings of the 1984 Harlaxton Symposium.* Nottingham: University of Nottingham Press, for Harlaxton College, 1985, pp. 31–44 and plates.

Campbell, B. M. S. "Agricultural Progress in Medieval England: Some Evidence from Eastern Norfolk." *EcHR,* 2d ser. 36 (1983):26–46.

Cannon, Henry Lewis. "The Battle of Sandwich and Eustace the Monk." *EHR* 27 (1912):649–70.

Carlyle, Thomas. *Past and Present.* London, 1843.

Carus-Wilson, Eleanora M. "An Industrial Revolution of the Thirteenth Century." *EcHR* 11 (1941):39–60.

———. "The English Cloth Industry in the Late Twelfth and Early Thirteenth Centuries." *EcHR* 14 (1944–1945):32–50.

Cheney, Christopher R. *From Becket to Langton: English Church Government, 1170–1213.* Manchester, 1956.

———. *Hubert Walter.* London: Thomas Nelson and Sons, 1967.

———. "English Cistercian Libraries: The First Century." In Christopher R. Cheney, *Medieval Texts and Studies.* Oxford: Clarendon Press, 1973, pp. 328–49.

Cheney, Mary G. "The Decretal of Pope Celestine III on Tithes of

Windmills, JL 17620." *Bulletin of Medieval Canon Law*, n.s. 1 (1971):63–66.

Chibnall, A. C. *Beyond Sherington*. Chichester: Phillimore and Co., 1979.

Cipolla, Carlo M. *Before the Industrial Revolution*. New York: Norton, 1976.

————, ed. *The Fontana Economic History of Europe*. Vol. 1, *The Middle Ages*. Glasgow, 1972.

Clay, Charles T. "The Family of Amundeville." *Lincolnshire Architectural and Archaeological Society Reports and Papers*, n.s. 3 (1945):109–37.

————. "Notes on the Family of Amundeville." *Archaeologia Aeliana*, 4th ser. 24 (1946):60–70.

Coelho, Antonia Borghes. *Portugal na Espanha Arabe*. Vol. 4. Lisbon: Sears Nova, 1975.

Colvin, H. M. "A List of the Archbishop of Canterbury's Tenants by Knight Service in the Reign of Henry II." In Francis R. H. Du Boulay, ed., *Documents Illustrative of Medieval Kentish Society*. Kent Archaeological Society Records Branch 18 (1964):1–40.

Constable, Giles. *Monastic Tithes from Their Origins to the Twelfth Century*. Cambridge: Cambridge University Press, 1964.

Daumas, Maurice. *A History of Technology and Invention*. Translated by Eileen B. Hennessy. 2 vols. New York: Crown Publishers, 1969.

Davenport, Frances Gardiner. *The Economic Development of a Norfolk Manor, 1086–1565*. Cambridge, England, 1906.

Davis, G. R. C. *Medieval Cartularies of Great Britain: A Short Catalogue*. London, 1958.

Delisle, Leopold. *Etudes sur la condition de la classe agricole et l'état d'agriculture en Normandie au moyen âge*. Paris, 1851.

————. "On the Origin of Windmills in Normandy and England." *Journal of the British Archaeological Association* 6 (1851):403–6.

Dodwell, Barbara. "Holdings and Inheritance in Medieval East Anglia." *EcHR*, 2d ser. 20 (1967):53–66.

Donkin, R. A. *The Cistercians: Studies in the Geography of Medieval England and Wales*. Toronto: Pontifical Institute of Medieval Studies, 1978.

Du Boulay, Francis R. H. *The Lordship of Canterbury*. London, 1966.

Duby, Georges. "Medieval Agriculture." In Cipolla, *Fontana Economic History*, 1:175–220.

Duggan, Charles. *Twelfth-Century Decretal Collections and Their Importance in English History*. London, 1963.

―――. "Primitive Decretal Collections in the British Museum." *Studies in Church History* 1 (1964):132–44.

Duncombe, John, and Battely, Nicholas. *The History and Antiquities of the Three Episcopal Hospitals at or near Canterbury: viz., St. Nicholas, at Harbledown; St. John's, Northgate; and St. Thomas, of Eastbridge. Bibliotheca Topographica Britannica*, vol. 1 (no. 30), 1785.

Earnshaw, J. R. "The Site of a Medieval Post Mill and Prehistoric Site at Bridlington." *Yorkshire Archaeological Journal* 45 (1973):19–40.

Ekwall, Eilert. *The Concise Oxford Dictionary of English Place-Names*. 4th ed. Oxford: Clarendon Press, 1960.

Elliott, J. S. "Windmills of Bedfordshire, Past and Present." *Bedfordshire Historical Record Society Octavo Publications* 14 (1931):3–50.

English, Barbara. *The Lords of Holderness, 1086–1260*. Oxford: Oxford University Press, for the University of Hull, 1979.

Eyton, R. W. *Court, Household, and Itinerary of King Henry II*. London, 1878.

Farmer, David Hugh. *The Oxford Dictionary of Saints*. Oxford: Clarendon Press, 1978.

Farmer, D. L. "Some Price Fluctuations in Angevin England." *EcHR*, 2d ser. 9 (1956–1957):34–43.

Farnham, George A. *Leicestershire Medieval Village Notes*. 6 vols. Leicester, 1929–1933.

Farrer, William. *Feudal Cambridgeshire*. Cambridge, England, 1920.

Faull, M. L., and Moorhouse, S. A. *West Yorkshire: An Archaeological Survey*. Chap. 2, "Cornmills." West Yorkshire Metropolitan County Council, 1981, pp. 701–23.

Fleming, Stuart. "Gallic Waterpower: The Mills of Barbegal." *Archaeology* 36 (Nov.–Dec. 1983):68–69, 77.

Freese, Stanley. *Windmills and Millwrighting*. Cambridge: Cambridge University Press, 1957.

Freese, Stanley, and Hopkins, R. T. *In Search of English Windmills*. London, 1931.

Gillett, Edward. *A History of Grimsby*. Oxford: Oxford University Press, for the University of Hull, 1970.

Gimpel, Jean. *The Medieval Machine: The Industrial Revolution of the Middle Ages*. New York: Holt, Rinehart and Winston, 1976.

Giraud, C. J. B. *Essai sur l'histoire du droit français au moyen âge.* Vol. 2. Paris, 1846.

Glick, Thomas F. *Islamic and Christian Spain in the Early Middle Ages.* Princeton, N.J.: Princeton University Press, 1979.

Gottfried, Robert S. *Bury St. Edmunds and the Urban Crisis, 1290–1539.* Princeton, N.J.: Princeton University Press, 1982.

Gras, Norman Scott Brien. *The Evolution of the English Corn Market, from the Twelfth to the Eighteenth Centuries.* Cambridge, Mass.: Harvard University Press, 1915. Reprint. New York: Russell and Russell, 1967.

Greenway, Diana E. *John Le Neve's Fasti Ecclesiae Anglicanae, 1066–1300.* Vol. 1 (*St. Paul's, London*); vol. 2 (*Monastic Cathedrals*); vol. 3 (*Lincoln*). London: Institute of Historical Research, 1968, 1971, 1977.

Hall, D. M. "A Thirteenth-Century Windmill Site: Strixton, Northamptonshire." *Bedfordshire Archaeological Journal* 8 (1973):109–18.

Harvey, P. D. A. "The English Inflation of 1180–1220." *Past and Present* no. 61 (Nov. 1973), pp. 3–30.

Hassall, W. O. "The Family of Jordan de Briset." *Genealogists' Magazine* 9 (1946):21–23.

Hatcher, John. "English Serfdom and Villeinage: Towards a Reassessment." *Past and Present* no. 90 (Feb. 1981):3–39.

Heales, Alfred. "Tandridge Priory and the Austin Canons." *Surrey Archaeological Collections* 9 (1888):19–156.

Hill, Derek Ingram. *Eastbridge Hospital and the Ancient Almshouses of Canterbury.* Canterbury: A. J. Snowdon, Ltd.

Hill, Frances. *Medieval Lincoln.* Cambridge, England, 1965.

Hilton, R. H. *Peasants, Knights, and Heretics.* Cambridge: Cambridge University Press, 1976.

———. "Freedom and Villeinage in England." *Past and Present* 31 (July 1965):3–19.

Holden, E. W. "Ancient Windmill Site at Glynde." *Sussex Archaeological Collections* 113 (1975):191–92.

Horrent, Jules. "La chanson du pèlerinage de Charlemagne et la réalité historique contemporaine." In *Mélanges de langue et de littérature du moyen âge et de la Renaissance offerts à Jean Frappier.* Geneva, 1970, 1:411–17.

Hoskins, W. G. *The Midland Peasant: The Economic and Social History of a Leicestershire Village*. London, 1957.

Hoving, Thomas. *King of the Confessors*. New York: Simon and Schuster, 1981.

Hughes, H. C. "Windmills in Cambridgeshire and the Isle of Ely." *Cambridge Antiquarian Society Proceedings* 31 (1922):17–30.

Hurry, Jamieson B. *Reading Abbey*. London, 1901.

Hussey, Arthur. "The Hospitals of Kent." *The Antiquary* 45 (1909): 414, 447; 46 (1910):139; 47 (1911):15–19, 97–101.

Hynes, Alice R. "The Mill in England, 1066–1307: The Influence of Technology on Society." M.A. thesis, Birbeck College, University of London, 1980.

Jarvis, P. S. *Stability in Windmills and the Sunk Post Mill*. Reading, England: International Molinological Society, 1981–1982.

Jenkinson, Hilary. "William Cade, A Financier of the Twelfth Century." *EHR* 110 (1913):209–27.

Kealey, Edward J. *Roger of Salisbury, Viceroy of England*. Berkeley and Los Angeles: University of California Press, 1972.

———. *Medieval Medicus: A Social History of Anglo-Norman Medicine*. Baltimore: The Johns Hopkins University Press, 1981.

———. "King Stephen: Government and Anarchy." *Albion* 6 (1974): 201–17.

———. "Anglo-Norman Royal Servants and the Public Welfare." *Albion* 10 (1978):341–51.

———. "Recent Writing about Anglo-Norman England." *British Studies Monitor* 9 (1979):3–22.

———. "Schools, Teachers, and Historians of the Reign of Henry I, 1100–1135." In Frederick M. Schweitzer, ed., *Festschrift in Honor of Brother Casimir Gabriel Costello, F.S.C.* New York: Manhattan College, 1983, pp. 1–22.

———. "England's Earliest Women Doctors." *Journal of the History of Medicine and Allied Sciences* 40 (1985):473–77.

———. "Hospitals and Poor Relief." In Joseph R. Strayer, ed., *Dictionary of the Middle Ages*. New York: Charles Scribner's Sons, 1985, 6:292–97.

Keil, Ian. "Building a Post Windmill in 1342." *TNS* 34 (1961–1962): 151–54.

King, Edmund. "The Origins of the Wake Family: The Early History of the Barony of Bourne in Lincolnshire." *Northamptonshire Past and Present* 5 (1975):166–76.

Knowles, David. *The Monastic Order in England*. Cambridge: Cambridge University Press, 1940.

Knowles, David; Brooke, C. N. L.; and London, Vera C. M. *The Heads of Religious Houses: England and Wales, 940–1216*. Cambridge: Cambridge University Press, 1972.

Knowles, David, and Hadcock, R. Neville. *Medieval Religious Houses: England and Wales*. New York: St. Martin's Press, 1971.

Kuttner, Stephen, and Rathbone, Eleanor. "Anglo-Norman Canonists of the Twelfth Century: An Introductory Study." *Traditio* 7 (1949–1951):279–358.

Lally, J. E. "Secular Patronage at the Court of King Henry II." *Bulletin of the Institute of Historical Research* 49 (1976):159–84.

Landon, Lionel. "The Bainard Family in Norfolk." *Norfolk Archaeology* 22 (1926):209 20.

Langdon, John. "Horse Hauling: A Revolution in Vehicle Transport in Twelfth- and Thirteenth-Century England." *Past and Present* no. 103 (May 1984), pp. 37–66.

Latham, R. A. *Revised Medieval Latin Word List from British and Irish Sources*. London: The British Academy, 1965.

Legge, M. Dominica. *Anglo-Norman Literature and Its Background*. Oxford: Clarendon Press, 1963

Lennard, Reginald. *Rural England, 1086–1135: A Study of Social and Agricultural Conditions*. Oxford: Oxford University Press, 1959.

———. "An Early Fulling Mill: A Note." *EcHR* 17 (1947):150.

———. "Early English Fulling Mills: Additional Examples." *EcHR*, 2d ser. 3 (1950):342–44.

Lewis, John. *The History and Antiquities, Ecclesiastical As Well As Civil, of the Isle of Thanet, Kent*. London, 1936.

Lloyd, John E. "The Welsh Chroniclers." *Proceedings of the British Academy* 14 (1928):369–91.

Long, W. Harwood. "The Low Yields of Corn in Medieval England." *EcHR*, 2d ser. 32 (1979):459–69.

Loomis, Laura Hibbard. *Adventures in the Middle Ages*. New York: Burt Franklin, 1962.

Lowndes, G. Alan. "History of the Priory at Hatfield Regis, Alias Hatfield Broad Oak." *Transactions of the Essex Archaeological Society*, n.s. 2 (1884):117–52.

Luchaire, Achille. *Social France at the Time of Philip Augustus*. Translated by Edward B. Kirkbiel. New York: Frederick Ungar, 1912.

MacGregor, Anne, and MacGregor, Scott. *Windmills*. London: Pepper Press, 1982.

McLachlan, Elizabeth Parker. "In the Wake of the Bury Bible: Followers of Master Hugh at Bury St. Edmunds." *Journal of the Warburg and Courtauld Institutes* 42 (1979):216–24.

Marques, Antonio Henrique R. de Oliveira. *Introdução à história da agricultura em Portugal: A questão cerealifera durante a Idade Média*. Lisbon, 1968; 3d ed., 1978.

Mayr-Harting, Henry. *The Bishops of Chichester, 1075–1207*. Chichester City Council, 1963.

Michinton, Walter. "Wind Power." *History Today* 30 (March 1980):3–36.

Miller, Edward. *The Abbey and Bishopric of Ely*. Cambridge: Cambridge University Press, 1951.

———. "England in the Twelfth and Thirteenth Centuries: An Economic Contrast?" *EcHR*, 2d ser. 24 (1971):1–14.

Miller, Edward, and Hatcher, John. *Medieval England: Rural Society and Economic Change*. New York: Longman, 1978.

Mortimer, Richard. "Religious and Secular Motives for Some English Monastic Foundations." *Studies in Church History* 15 (1978):77–85.

———. "The Beginnings of the Honour of Clare." *Proceedings of the Battle Abbey Conference on Anglo-Norman Studies* 3 (1980):106–99.

———. "The Family of Rannulf de Glanville." *Bulletin of the Institute of Historical Research* 54 (1981):1–16.

Needham, Joseph. *Science and Civilization in China*. Vol. 1. Cambridge: Cambridge University Press, 1954.

Nichols, John. *The History and Antiquities of the County of Leicester*. 4 vols. in 8 pts. London, 1795–1815.

Nichols, John, et al. *Bibliotheca Topographica Britannica*. 10 vols. London, 1780–1800.

Norgate, Kate. "The *Itinerarium Peregrinorum* and the *Song of Ambrose*." *EHR* 25 (1910):523–47.

Ovitt, George, Jr. "The Status of Mechanical Arts in Medieval Classifications of Learning." *Viator* 14 (1983):89–106.

Page, Frances M. *The Estates of Crowland Abbey*. Cambridge: Cambridge University Press, 1934.

Painter, Sidney. *Studies in the History of the English Feudal Barony*. Baltimore: The Johns Hopkins University Press, 1943.

Pearce, S. V. "A Medieval Windmill, Honey Hill, Dogsthorpe." *Proceedings of the Cambridge Antiquarian Society* 49 (1966):95–104.

Peckham, W. D. "The Architectural History of Amberley Castle." *Sussex Archaeological Collections* 62 (1921):21–63.

Pérés, Henri. *La poésie Andalouse en Arabe classique*. Paris, 1937, 1953.

Posnansky, M. "The Lamport Post Mill." *Journal of the Northamptonshire Natural History Society and Field Club* 33 (1956):66–79.

Postan, M. M. *The Medieval Economy and Society: An Economic History of Britain, 1100–1500*. Berkeley and Los Angeles: University of California Press, 1972.

Pritchard, Violet. *English Medieval Graffiti*. Cambridge: Cambridge University Press, 1967.

Rahtz, Philip, and Bullough, Daniel. "The Parts of an Anglo-Saxon Mill." *Anglo-Saxon England* 6 (1977):15–37.

Randall, Lillian M. C. *Images in the Margins of Gothic Manuscripts*. Berkeley and Los Angeles: University of California Press, 1966.

Reynolds, John. *Windmills and Watermills*. London: Hugh Evelyn, 1970. New York: Praeger, 1970

Reynolds, Terry S. *Stronger than a Hundred Men: A History of the Vertical Water Wheel*. Baltimore: The Johns Hopkins University Press, 1983.

Richardson, Henry G. *The English Jewry under the Angevin Kings*. London, 1960.

Rigold, S. E. "Two Kentish Hospitals Re-examined: St. Mary Ospring and SS. Stephen and Thomas, New Romney." *Archaeologia Cantiana* 79 (1964):31–69.

Riley, Henry Thomas. "The History and Charters of Ingulfus Considered." *Archaeological Journal* 19 (1862):32–49, 114–33.

Riley-Smith, Jonathan. *The Knights of St. John in Jerusalem and Cyprus, 1050–1310*. New York: St. Martin's Press, 1967.

Roehl, Richard. "Patterns and Structures of Demand, 1100–1500." In Cipolla, *Fontana Economic History*, 1:106–42.

Rollason, D. W. "The Cults of Murdered Royal Saints in Anglo-Saxon England." *Anglo-Saxon England* 11 (1983):1–22.

Roth, Cecil. *A History of the Jews in England.* 3d ed. Oxford: Oxford University Press, 1964.

Round, John Horace. "The Foundation of the Priories of Saint Mary and Saint John, Clerkenwell." *Archaeologia* 56 (1899):223–28.

———. "The Fitz Walter Pedigree." *Essex Archaeological Society Transactions*, n.s. 7 (1900):329–30.

———. "The Founder of Stanegate Priory." *Essex Archaeological Society Transactions*, n.s. 14 (1918):218–26.

———. "The Early Sheriffs of Norfolk." *EHR* 35 (1920):481–96.

Rowley, Trevor. *The Norman Heritage, 1066–1200.* Boston: Routledge and Kegan Paul, 1983.

Russell, J. C. "Population in Europe, 1000–1500." In Cipolla, *Fontana Economic History*, 1:25–70.

Runnels, Curtis N. "Milling in Ancient Greece." *Archaeology* 36 (Nov.–Dec. 1983):62–63, 75.

Salmon, John. "The Windmill in English Medieval Art." *Journal of the Archaeological Association*, 3d ser. 6 (1941):88–102.

Sanders, Ivor J. *English Baronies: A Study of Their Origin and Descent, 1086–1327.* Oxford: Clarendon Press, 1960.

Sawyer, Peter H. *Anglo-Saxon Charters: An Annotated List and Bibliography.* London: Royal Historical Society, 1968.

Schlauch, Margaret. "The Palace of Hugon de Constantinople." *Speculum* 7 (1932):500–514.

Searle, W. G. *Ingulf and the Historia Croylandensis.* Cambridge, England: Cambridge Antiquarian Society, 1894.

Smith, R. A. L. *Canterbury Cathedral Priory: A Study in Monastic Administration.* Cambridge: Cambridge University Press, 1969.

Smith, Terence Paul. "Biggleswade Windmill." *Bedfordshire Archaeological Journal* 8 (1973):136–37.

———. "The English Medieval Windmill." *History Today* 28 (1978): 256–63.

———. "Windmill Graffiti at Saint Giles' Church, Tottenhoe." *Bedfordshire Archaeological Journal* 14 (1980):104–6.

Somner, William. *Antiquities of Canterbury.* Canterbury, 1640; 2d ed. (2 pts.), 1703.

Spufford, Margaret. *A Cambridgeshire Community: Chippenham from Settlement to Enclosure.* Leicester University Occasional Papers 20 (1965).

Stenton, Doris M. *English Justice between the Norman Conquest and the Great Charter, 1066–1215.* Memoirs of the American Philosophical Society 60 (1964).

Stokes, H. P. "The Old Mills of Cambridge." *Cambridge Antiquarian Society Proceedings* 14 (1910):180–233.

Strutt, Joseph. *Honda Angel-ynnan, or A Complete View of the Manners, Customs, Arms, Habits, etc. of the Inhabitants of England.* 2 vols. London, 1775–1776.

Sypherd, Wilbur Owen. *Studies in Chaucer's House of Fame.* Chaucer Society, 1907.

Talbot, C. H., and Hammond, E. A. *The Medical Practitioners in Medieval England: A Biographical Register.* London: Wellcome Historical Medical Library, 1965.

Tann, Jennifer. "Multiple Mills." *Medieval Archaeology* 11 (1967):253–55.

Thompson, Alexander Hamilton. *The English Clergy and Their Organization in the Later Middle Ages.* Oxford: Clarendon Press, 1947.

Thrupp, Sylvia L. "Medieval Industry, 1000–1500." In Cipolla, *Fontana Economic History*, 1:221–73.

Thurston, Herbert, and Attwater, Donald. *Butler's Lives of the Saints.* 4 vols. New York: P. J. Kenedy and Sons, 1956.

Titley, Arthur. "Notes on Old Windmills." *TNS* 3 (1924):41–51.

Titow, J. Z. *English Rural Society, 1200–1350.* London: George Allen and Unwin, Ltd., 1969.

Turner, Ralph V. *The King and His Courts.* Ithaca, N.Y.: Cornell University Press, 1968.

———. *The English Judiciary in the Age of Glanville and Bracton, c. 1176–1239.* Cambridge: At the University Press, 1985.

———. "William de Forz, Count of Aumale: An Early-Thirteenth-Century English Baron." *Proceedings of the American Philosophical Society* 115, no. 3 (June 1971):221–49.

Urry, William J. *Canterbury under the Angevin Kings.* London, 1967.

———. "Two Notes on Guernes de Ponte-Sainte-Maxence: *Vie de Saint Thomas.*" *Archaeologia Cantiana* 66 (1953–1954):92–97.

————. "Saint Anselm and His Cult at Canterbury." *Spicilegium Beccense*. Congrès international du IX^e centenaire de l'arrivée d'Anselm au Bec. Le Bec-Hellouin and Paris, 1959, 1:571–93.

Van Engen, John. "Theophilus Presbyter and Rupert of Deutz: The Manual Arts and Benedictine Theology." *Viator* 11 (1980):147–64.

*V.C.H.* Doubleday, H. Arthur; Page, William; Salzman, Louis F.; and Pugh, Ralph B., eds. *The Victoria History of the Counties of England.* London and Westminster, 1900–1934. London: Institute of Historical Research, 1935–.

Vowles, Hugh P. "An Inquiry into Origins of the Windmill." *TNS* 11 (1930):1–14.

————. "Early Evolution of Power Engineering." *ISIS* 17 (1932):412–20.

Wailes, Rex. *Windmills in England*. London, 1948.

————. *The English Windmill*. London: Routledge and Kegan Paul, 1954, 1968 (reprint).

————. *Source Book of Windmills and Watermills*. London: Ward and Lock, 1979.

————. "Suffolk Windmills; Part I, Post Mills." *TNS* 22 (1941–1942):41–65.

————. "The Windmills of Cambridge." *TNS* 27 (1949–1951):97–121.

————. "Essex Windmills." *TNS* 31 (1957–1958):153–82.

Ward, J. C. "Fashions in Monastic Endowment: The Foundations of the Clare Family, 1066–1314." *JEH* 32 (1981):427–51.

White, Lynn, Jr. *Medieval Technology and Social Change*. New York: Oxford University Press, 1962.

————. *Medieval Religion and Technology*. Berkeley and Los Angeles: University of California Press, 1978.

————. "Eilmer of Malmesbury, An Eleventh-Century Aviator: A Case Study of Technological Innovation, Its Context and Tradition." *Technology and Culture* 2 (1961):97–111.

————. "Medieval Uses of Air." *Scientific American* 223, no. 2 (August 1970):92–100.

————. "The Expansion of Technology, 500–1500." In Cipolla, *Fontana Economic History*, 1:143–74.

Wilson, S. Gordon. *With the Pilgrims to Canterbury and the History of the Hospital of Saint Thomas*. Canterbury, 1934.

Young, Charles R. *Hubert Walter, Lord of Canterbury and Lord of England*. Durham, N.C.: Duke University Press, 1968.

Zeepvat, R. J. "Post Mills and Archaeology." *Current Archaeology* 6 (1980):375–77.

# Index

Designer: Wolfgang Lederer
Compositor: Wilsted & Taylor
Text: 12 × 14½ Garamond
Display: Garamond
Printer: Malloy Lithographing
Binder: John H. Dekker & Sons